Vedic Mathematics and the Nature of Time

Vedic Mathematics and the Nature of Time

Three Essays by

VASYL SEMENOV, PHD

Acoustics and Computational Mathematics
Ukrainian National Academy of Sciences

Translated by the Author

rlta

RLTA PRESS
Alachua, Florida

The publishers of this volume would like to offer their profound appreciation to the Bhaktivedanta Book Trust trustees whose dedicated support of the activities of the Bhaktivedanta Institute for Higher Studies facilitated the production of this volume.

First English Edition

ISBN: 978-1-959829-05-8

Editorial Team:
Prishni Sutton
Krishna Kripa
Vallabha Caitanya
Zoya Komissar
Vraja Kishore

Ukrainian edition of
"Vedic Mathematics" and "Nature of Time"

"Artificial Intelligence: Can a Computer Possess Consciousness?"

For more information about the contents of this work,
please contact the author at: dvija.govinda.ids@gmail.com

For more information about the activities of the
Bhaktivedanta Institute for Higher Studies
Please see: www.bihstudies.org

Dedicated to A. C. Bhaktivedanta Swami Prabhupāda

jñānaṁ te 'haṁ sa-vijñānam
idaṁ vakṣyāmy aśeṣataḥ
yaj jñātvā neha bhūyo 'nyaj
jñātavyam avaśiṣyate

I shall now declare unto you in full this knowledge, both phenomenal and numinous. This being known, nothing further shall remain for you to know.

—*Bhagavad-gītā As It Is,* 7.2

Contents

Foreword *ix*
Introduction to the English Edition *xi*

PART ONE: Vedic Mathematics **1**

Preface 3
Introduction 4
1. Vedic Knowledge 6
2. Descending Method of Learning 7
3. Decimal Numeral System 10
4. Vedic Arithmetic 14
5. Vedic Geometry 24
6. Vedic Trigonometry 31
7. Vedic Combinatorics 35
8. On the Nature of Mathematical Creativity 44
9. Brief Information about Mathematicians of the Vedic Tradition 49

Appendix. Vedic Units of Measure for Distance and Time 53

PART TWO: Nature of Time **55**

Introduction 57
1. A Brief Historical Review 60
2. Vedic Concept of Time 66
3. The Creator's Plan 69
4. The Role of Time in the Creation of the Universe 72
5. Destructive Action of Time 74
6. The Measurement of Time 76
7. Measurement of Long Periods of Time 80
8. Our Position in Vedic Chronology 82
9. Time at Different Planetary Levels 84

10. Time in the Spiritual World 87
11. The Cyclical Nature of Time and a Look at World History 90
12. Vedic Calendar 94
13. Time Travel 99
14. How To Overcome the Influence of Time 103
15. Is Everything Predetermined? 105
Conclusions 109

PART THREE: Artificial Intelligence: Can a Computer Possess Consciousness? 111

Foreword

This is a highly readable and informative book explaining Hindu concepts of time, based primarily on the translation and commentaries of the *Bhāgavata Purāṇa* by A. C. Bhaktivedanta Swami Prabhupāda, which represents the philosophical point of view of the International Society for Krishna Consciousness (ISKCON). It is quite thorough, explaining not just the way time has been measured, but also its relationship with other aspects of creation, sustenance, and destruction of the world, spirituality, etc. It is amazing how my ancestors could conceive scales of time from the time it takes for the sun to traverse the diameter of a molecule to the entire lifetime of the universe. The essay is a short one, which should inspire further research. For example, what instruments were used to measure times at these varied scales? Are there any connections with the various time scales with astronomical phenomena such as the precession of the earth, time it takes the sun to revolve around the galaxy, etc., and viewpoints other than those of ISKCON alone? I admire the efforts of the author, especially given the fact that he worked on writing such a book in the middle of war with bombs bursting all around him. In my opinion this book would make a great addition to anyone's library.

— Hrushikesh Mhaskar
Distinguished Research Professor
Institute of Mathematical Sciences
Claremont Graduate University, California, USA
(*Opinions are personal*)

Introduction to the English Edition

Vasyl Semenov (Dvija Govinda) became involved with the Bhaktivedanta Institute for Higher Studies (BIHS) in 2019 when he presented at the "Consciousness in Science" conference, co-sponsored by the BI and the University of Florida Center for the Study of Hindu Traditions (CHiTra). His paper, "Comparative Analysis of Cosmological Concepts of *Śrīmad-Bhāgavatam* with Modern Astrophysics,"[1] was co-authored with his wife, Evgeniya Semenova (Sukumārī Sundarī), who holds a PhD in Mathematics. Semenov earned his first PhD in Acoustics in 2004 from the Institute of Hydromechanics of the National Academy of Sciences of Ukraine (NASU), and a second PhD in Computational Mathematics in 2020 at the V. M. Glushkov Institute of Cybernetics of NASU (2020).

Semenov has presented a number of his interests in Purāṇic cosmology during three successive BI Cosmology Workshops.[2] He offered an overview, "What is the *Brahmāṇḍa*?", at the November 2023 BI conference, "Cosmology of the *Bhāgavata Purāṇa*: Current Research on History, Philosophy, and Science."

In 2021 Semenov published two monographs – one on Vedic mathematics integral to the Sanskrit tradition of observable astronomy (*jyotiṣa*), and a second on the nature of time as a central concern in cosmology. This publication includes these two essays translated into English by the author, along with a short article on artificial intelligence published in 2024.

Section One, "Vedic Mathematics," examines mathematics as an

---

1 Vasyl Semenov and Evgeniya Semenova, "Comparative Analysis of Cosmological Concepts of *Śrīmad-Bhāgavatam* with Modern Astrophysics," *2019 Consciousness in Science Conference: Partial Collection of Abstracts and Papers* (Alachua, FL: IVS Press, 2023), 110–12.

2 "A Proportionality Approach to the Relevance of 'Height' in the Purāṇic Cosmos," *Pūrva-pakṣa: Fine Tuning Opposing Views*, Vol. 1, (Alachua, FL: IVS Press, 2023), 21–37; and "Vertical Dimension and Size of the Purāṇic Universe: Existing Interpretations and New Insights," *Pūrva-pakṣa: Fine Tuning Opposing Views*, Vol. 2, (Alachua, FL: IVS Press, 2024), 133–47.

integral component of Vedic knowledge systems such as phonetics, grammar, etymology, metrics, astronomy, and astrology. Kim Plofker, the American historian of mathematics, describes in her book, *Mathematics in India*, a number of noteworthy examples of the unique role mathematics plays within the intellectual development of the Indian subcontinent:

> Aryabhata's decimal arithmetic [c. fifth century CE] is not associated with Greek [mathematics, similarly] ... Madhava's power series for trigonometric functions [c. fourteenth century] predate by centuries Newton's and Leibniz's versions of them.
>
> Vedic texts clearly indicate a long-standing tradition of decimal numeration and a deep fascination with various concepts of finite and infinite quantities and their significance in the cosmos.
>
> [*Jyotiṣa*] treatises ... sometimes reveal intriguing and creative mathematical approaches, including some of the earliest known uses in Indian texts of iterative approximation techniques.[3]

Semenov's treatise further explores these and other areas of mathematics that within this historical culture connect material and spiritual perspectives in a manner that encourages self-realization.

Section Two, "The Nature of Time," offers a comparative examination of Vedic and contemporary analyses on the nature of time. It considers questions of the influence of time on external reality, as well as time scales for measuring micro- and macro-processes within the universe, questions of determinism, and the possibility of time travel. Semenov's essay considers a wide range of interests identified with Vedic knowledge systems involving general laws of creation and maintenance that appear foundational to the cosmos.

Part Three, "Artificial Intelligence: Can a Computer Possess Consciousness?" offers a brief historical overview on the subject. Semenov examines perspectives on artificial intelligence from scholars such as

---

3 Kim Plofker, *Mathematics in India* (Princeton: Princeton University Press, 2009), pages 4, 16, 88.

Nobel Laureate Sir Roger Penrose, along with teachings described in the *Bhagavad-gītā*.

The intent of this English language edition is to help enrich the study of the exoteric cosmological accounts described within the Vedic Sanskrit tradition of *jyotiṣa* mathematical astronomy, along with the esoteric cosmological narratives located within *Purāṇas* such as the Fifth Canto of the *Śrīmad-Bhāgavatam*, while considering potential relationships between the two.

PART ONE

Vedic Mathematics

Preface

Mathematics is considered to be the queen of all knowledge. As the language of modern science, it has allowed humanity to make significant progress in the study of the laws of nature. It is interesting that the prototypes of many achievements of modern mathematics are contained in the manuscripts of ancient cultures, such as the Vedic civilization, which was spread throughout the territory of modern India and beyond. The practical and holistic nature of Vedic mathematics made it an important part of the *Vedas*, one of the oldest monuments of human culture and science. It should be noted that knowledge of history also contributes to the progress of specific mathematical disciplines. Thus, information from Vedic mathematics has found application in the construction of modern high-speed multipliers on digital microcircuits, and is also used in a number of countries (USA, India) as part of the school mathematics curriculum. I believe that this book has scientific and practical value and may be of interest to readers who are interested in Vedic knowledge and the history of mathematics.

—Valeriy K. Zadiraka, Ph.D., D.Sc.,
Academician of National Academy of Sciences of Ukraine

Introduction

"Mathematics is the queen of the sciences."
Carl Friedrich Gauss (1777–1855)

What is mathematics? The classic definition is "the science that studies quantitative relations and spatial forms of the surrounding world." Throughout the history of mankind, mathematics has played the role of the "Queen of Sciences," since other scientific disciplines largely rely upon it. At present, nearly all authoritative scientific theories use the language of mathematics to express their ideas.

Nowadays, mathematics has literally become the language of modern science, which has allowed humanity to achieve certain technological advances. Roger Penrose, one of the most prominent scientists of our time and the 2020 Nobel laureate in physics, says that "people are often puzzled that something abstract like mathematics can really describe reality." According to him, it is impossible to understand atomic particles and structures such as gluons and electrons except through mathematics.

Albert Einstein once remarked: "How can it be that mathematics, being ultimately a product of human thought, independent of experience, is so excellently suited to objects of reality?" And indeed, since ancient times, humanity has debated whether mathematics was *discovered* or *invented*. In other words, have we created mathematical concepts to help us understand the universe around us, or is mathematics the native language of the universe itself? Are numbers, polygons, and equations real, or are they some kind of theoretical ideal? The ancient Pythagoreans considered numbers to be both living things and universal principles. Plato argued that mathematical concepts are concrete and as real as the universe itself, no matter what we know about them. Euclid, "the father of geometry," believed that nature itself is the physical manifestation of mathematical laws.

In 1960, Nobel laureate in physics Eugene Wigner introduced the concept of the "incomprehensible effectiveness of mathematics," strongly promoting the idea that mathematics has objective existence and was discovered by humans.[1] Wigner (who, by the way, studied the Vedic scriptures in detail) pointed out that many mathematical theories, developed in a vacuum (i.e., by themselves), often appeared unrelated to the description of any physical phenomena. But decades or even centuries later, they turned out to be the basis needed to explain how the universe works.

Roger Penrose, already mentioned above, believes that mathematics has its own independent, "Platonic" existence. He compares mathematics to geology or archeology, where you explore really beautiful things that have actually existed for centuries, and you discover them for the first time. Also in his research, Penrose explores a concept of three worlds: physical, mental, and mathematical (Platonic). This is certainly reminiscent of the Vedic concepts: gross matter (physical), subtle energy (mental), and the spiritual world (the source of gross and subtle forms manifested in the material world).

---

1 Wigner, E. P. (1960). "The unreasonable effectiveness of mathematics in the natural sciences. Richard Courant lecture in mathematical sciences delivered at New York University, May 11, 1959". *Communications on Pure and Applied Mathematics*. 13 (1): 1–14.

1. Vedic Knowledge

"Access to the *Vedas* is the greatest privilege this century may claim over all previous centuries."
Oppenheimer (1904–1967)

The *Vedas* are the source of some of the oldest knowledge known to mankind. The word *veda*, derived from the Sanskrit root *vid*, literally means "knowledge." Vedic knowledge is also called *apauruṣeya*, which means that the *Vedas* are not an independent creation of the human mind. According to the *Vedas*, the material world around us passes through continual periodic cycles, lasting several billion years. Material and spiritual knowledge arises initially at the beginning of the creation of the universe, and then is transmitted along chains of disciplic succession.

Another name for Vedic knowledge is *śruti*, which means "heard." In previous centuries, people had a keen memory, and disciples were able to quickly memorize the knowledge received from their spiritual master. However, about 5000 years ago, Śrīla Vyāsadeva foresaw that the people of the age of Kali would not have such qualities. Therefore, he divided the Vedic scriptures into four *Vedas*, *Upaniṣads*, *Purāṇa*s, *Itihāsas*, and *Vedāṅgas*.[2]

Vedic mathematics is an internal component of Vedic knowledge. Although not being singled out as a separate, independent discipline, it nevertheless serves to link the various sections of the *Vedas* and helps them achieve their main goal: to lead a practitioner along the path of realizing the Absolute Truth. So, one of the abovementioned *Vedas,* the Vedāṅgas (literally "parts of the *Vedas*"), includes phonetics, grammar, etymology, metrics, astronomy, and the discipline of ritual performances. As we will see in the course of presenting this work, these knowledge systems are saturated with mathematics, which serve as the basis – the inner fabric or the bridge – connecting the material and spiritual aspects of knowledge.

---

2 A. C. Bhaktivedanta Swami Prabhupāda, *Śrīmad-Bhāgavatam* (SB), Canto 1 (Los Angeles, CA: The Bhaktivedanta Book Trust, 1980), 4.19–25.

2. Descending Method of Learning

"What's essential is invisible to the eye."
Antoine de Saint-Exupéry (1900–1944)

In scientific methodology, there are two processes of cognition – inductive (ascending) and deductive (descending). In the current era, the development of science is characterized by attempts to create a model of the universe by consistently generalizing experimental observations and the results of logical reasoning. Of course, this limits the possibilities of our understanding due to the imperfection of our senses and the devices created by them, as well as by the capabilities of the human mind. As Albert Einstein said, "We cannot solve our problems with the same thinking we used when we created them." As a result, a large number of different highly specialized scientific branches emerge, developing independently of each other. And so, modern science is trying to piece together a holistic picture by analyzing disparate parts of the universe.

In contrast, the descending, or top-down, method (called *avaroha-panthā* in Sanskrit) inherent to the Vedic approach involves taking knowledge from an authoritative source. This is also known as *śabda*, or the transmission of sound through vibration from a spiritual master (for more details on the four levels of sound vibration see Chapter 8: "On the Nature of Mathematical Creativity"). This approach tends to generate a more synthetic picture of the world, in which knowledge begins with a holistic vision, then descends to particular parts. In his book *Mechanics in Its Development* (1883) Ernst Mach writes: "The task of any science is the economy of thought and the economy of labor." The principle of *śabda*, or obtaining knowledge from authority, follows this principle. In *Science and Hypothesis*, a study written in 1902, Henri Poincaré makes a similar statement: "The importance of a fact is measured by its productivity, that is, by the amount of thought that it allows us to save."

For this reason, out of the cognitive methods, the *Vedas* distinguish *śabda* – the method of acquiring knowledge from an authoritative source through the chain of disciplic succession (*paramparā*). Of course, the Vedic approach does not reject the acquisition of knowledge by empirical means or by logical reasoning, but gives them a somewhat auxiliary role in relation to *śabda.*

Many modern scientists state emphatically that they cannot take on faith what cannot be directly verified by experiment or logical reasoning. However, it should be noted that negative numbers (–1, –2, –3...), known for millennia in Vedic culture, were considered fictitious in European science at least until the 16th century.[3] In addition, every scientist knows that mathematicians work with the imaginary number *i*, the square root of –1. Nowadays numerous practical technologies use algorithms that operate with imaginary numbers (e.g., digital communication devices, such as mobile phones, modems, etc.), and yet, the existence of such numbers is taken on faith, since they cannot be verified through any experiment. The same applies to quaternions and other abstract concepts of modern mathematics.

In the opinion of their contemporaries, the non-Euclidean geometries of Lobachevsky and Riemann, which appeared in the 19th century, had no practical value and were also devoid of common sense. Nevertheless, at the beginning of the 20th century, they found an application in Einstein's General Theory of Relativity, and later in many other areas as well. Likewise, the work on number theory by the British mathematician Gottfried Hardy, who boasted that none of his works would ever prove useful in describing any phenomena in the real world, helped the development of cryptography. Another part of his purely theoretical work became known as the famous Hardy-Weinberg law in genetics.

In this regard, Charles Townes, laureate of the 1964 Nobel Prize in Physics, noted that the faith of a scientist does not differ in essence from the faith of a religious person. This faith allows scientists to work for years to obtain a particular result. In addition, Townes argued that many important scientific discoveries, such as his invention of the laser, occur as a "flash," more like a religious revelation than an interpretation of data (see pages 39–41). Also, the revolutionary work of the famous Austrian mathematician, Kurt Gödel, revealed that in any consistent system of axioms there will be statements within the system

3 D.J. Struik, *A Concise History of Mathematics*, 1948.

that cannot be proved or refuted.[4] Of course, such statements *could* be added in to form a new system of axioms. But now in this new system (with the increased number of axioms), there will also exist unprovable and irrefutable statements! Thus, we cannot avoid accepting axioms that have to be taken on faith.

4 The popular name for Gödel's "Incompleteness theorem" is "the existence of God theorem."

3. Decimal Numeral System

"The appearance of the Indian number system in Europe
a thousand years earlier would have led to a significantly
higher level of development of modern science."
Carl Friedrich Gauss (1777–1855)

Perhaps the most famous contribution of Vedic mathematics is the use of the decimal number system. Indeed, the decimal number system was used in the *Vedas* more than five thousand years ago. Vedic mathematicians had developed a system of tens, hundreds, thousands, etc. Also in Vedic science, the concept of zero (*śunya*) was used.

To demonstrate that the decimal system was used in the *Vedas*, it is enough to consider the Sanskrit numeral *eka-daśa*, which means "eleven."[5] *Eka* means one and *daśa* means ten. Thus, to designate the number "eleven," a new designation was not introduced, but the digit 1 was transferred to the next position leftwards.

An example of this is found in *Śrīmad-Bhāgavatam* 1.3.16, which describes the eleventh *līlā-avatāra* of Kṛṣṇa in the form of Kūrma, a giant tortoise:

surāsurāṇām udadhiṁ
mathnatāṁ mandarācalam
dadhre kamaṭha-rūpeṇa
pṛṣṭha ekādaśe vibhuḥ

"The eleventh incarnation of the Lord took the form of a tortoise whose shell served as a pivot for the Mandarācala Hill, which was being used as a churning rod by the theists and atheists of the universe."

In 662, the Christian bishop of Syria, Severus Sebokht, wrote: "I will not touch upon the science of the Indians ... their number system that exceeds all descriptions. I just want to say that the count is made

5 Also, the Sanskrit word *ekādaśī* means the eleventh day of the lunar calendar.

using nine characters." The decimal number system came to Europe only in the 9th century AD due to the work of Al-Khwarizmi "On the Indian Account."[6] Leonardo of Pisa (Fibonacci) wrote about the decimal number system as the "Indian method" (*modus indorum*). In Europe, the decimal number system was finally established only about 500 years ago. The famous German mathematician C.F. Gauss (1777–1855) is credited with the assertion that the arrival of the Indian number system in Europe a millennium earlier would have led to a much more advanced level of modern science.

Table 1 shows the main Sanskrit numbers in the *devanāgarī* alphabet and their transliterations:[7]

TABLE 1

Sanskrit	Arabic	Transliteration
०	0	*śunya*
१	1	*ekam*
२	2	*dve*
३	3	*trini*
४	4	*catvari*
५	5	*pañca*
६	6	*ṣaṣ*
७	7	*sapta*
८	8	*aṣta*
९	9	*nava*
१०	10	*daśa*
११	11	*ekadaśa*
१२	12	*dvadaśa*
१३	13	*trayodaśa*

6 D. J. Struik, *A Concise History of Mathematics*, 1948.

7 Since the spelling of the numbers 1 to 4 depends on the gender used, the last column gives the order of the Sanskrit numbers in the neuter gender. Starting with the number 5, the spelling of the numerals in all three genders is the same.

Sanskrit	Arabic	Transliteration
१४	14	*caturdaśa*
१५	15	*pañcadaśa*
१६	16	*ṣodaśa*
१७	17	*saptadaśa*
१८	18	*aṣṭādaśa*
१९	19	*navadaśa* (*eka-una-vimśati*)[8]
२०	20	*vimśati*
३०	30	*trimśati*
१००	100	*śata*
१०१	101	*eka-śata* (*eka-uttara-śata*)
१०८	108	*aṣṭā-śata* (*aṣta-uttara-śata*)
२००	200	*dviśata*
३००	300	*triśata*
१०००	1 000	*sahasra*
१००००	10 000	*ayuta*
१०००००	100 000	*lakṣana*
१००००००	10 000 000	*koti*

Here is an example of the formation of a complex numeral: ३५६ – *ṣat panchaśat triśata* (literally "six + fifty + three hundred").

Let's take another example. Verse 5.21.7 of *Śrīmad-Bhāgavatam* states:

evaṁ nava koṭaya eka-pañcāśal-lakṣāṇi yojanānāṁ
mānasottara-giri-parivartanasyopadiśanti

"My dear King, as stated before, the learned say that the sun travels over all sides of Mānasottara Mountain in a circle whose length is 95,100,000 *yojanas*."

In this verse, the numeral *nava koṭi eka-pañcāsat-lakshaṇaḥ* literally

8 The word *una* in this context means "minus", that is, 19 = 20-1.

means "nine times ten million and fifty-one times one hundred thousand," that is, $9 \times 10{,}000{,}000 + 51 \times 100{,}000 = 95{,}100{,}000$.

In Vedic mathematics, special names are also provided for significantly larger numbers, shown in Table 2:

TABLE 2

Notation	Equivalent	Notation	Equivalent
ayuta	10^9	*hetu-hila*	10^{31}
niyuta	10^{11}	…	
kaṅkara	10^{13}	*niravadya*	10^{41}
…		…	
utsaṅgaha	10^{21}	*vibhūtaṃgamā*	10^{51}
…			

4. Vedic Arithmetic

As is known, the Vedic scriptures are often formulated in the form of extremely short statements, *sūtras*, which are "encrypted," that is, cannot be understood without a correct understanding of their context. Suffice it to recall the many different interpretations of the *ātmārāma* verse from *Śrīmad-Bhāgavatam* (1.7.10), which Lord Caitanya explained to Sārvabhauma Bhaṭṭācārya and Sanātana Gosvāmī, not to mention the many false interpretations that existed earlier. At the beginning of the twentieth century, the work of Tirthaji[9] gave a mathematical interpretation of the group of *sūtras* from *Atharva Veda*. Below we give his explanation of these *sūtras*.

1. *Ekādhikena pūrveṇa* (one more than the previous one)

This *sūtra* deals with the squaring (multiplying by itself) of numbers ending in "5". For example, we want to count the product 35 × 35.

According to this *sūtra*, we take the number formed by the digits preceding to the last 5 (i.e., 3), take the number that is greater than it by one (4), and multiply them by each other (3 × 4 = 12). On the right side, we attach "25" (no matter what number was squared) and get the answer: 1225.

Next example: 75 × 75.

1) Multiply 7 × (7 + 1) = 56.
2) Attach 25: 5625.

The same rule applies to arbitrarily large numbers (not only two-digit ones).

Consider multiplication 125 × 125.

1) Multiply 12 × (12 + 1) = 156.
2) Attach 25: 15,625.

---

9 Tirthaji J.S.S.B.K., *Vedic Mathematics*, Delhi, 1988.

Interestingly, this rule applies not only to squaring, but also to multiplying any numbers that differ only in the last digits, which add up to 10.

For example, consider the multiplication 53 × 57.

1) Multiply 5 × (5 + 1) = 30.
2) Add the product of the last digits (3 × 7 = 21). We get the answer: 3021.

Or, consider multiplication 102 × 108:

1) Multiply 10 × (10 + 1) = 110.
2) Add the product of the last digits (2 × 8 = 16).

Answer: 11,016.

Note. To verify calculations in Vedic mathematics the concept of a "root" was used. The root of a number is counted as the sum of all its digits. If the obtained value has more than one digit, then its digits are summed up again, etc.

The root of 102 is: 1 + 0 + 2 = 3.
The root of 108 is: 1 + 0 + 8 = 9.
The root of 191,983 is: 1 + 9 + 1 + 9 + 8 + 3 = 31, 3 + 1 = 4.

Let's look at how to use roots to verify computations.

The root of the product of two numbers is equal to the product of their roots. In the last example, the product of the roots of numbers 108 and 102 is 9 × 3 = 27 and 2 + 7 = 9. This means that the root of the answer (11,016) should be 9. Indeed, 1 + 1 + 0 + 1 + 6 = 9, i.e., the verification is successful.

The root of the sum of two numbers is equal to the sum of their roots. Let's check the result of calculating the sum 841 + 276 = 1117.

Root (841) = 8 + 4 + 1 = 13, 1 + 3 = 4.
Root (276) = 2 + 7 + 6 = 15, 1 + 5 = 6.

The sum of the roots is 4 + 6 = 10 and 1 + 0 = 1, while the root of the answer is 1 + 1 + 1 + 7 = 10, 1 + 0 = 1, that is, the verification is successful.

The root of the difference between two numbers is equal to the difference between their roots. If it turns out to be equal to zero or negative, then 9 should be added to it. Let's check the result of calculating the difference:

841 – 276 = 565.
Root (841) = 8 + 4 + 1 = 13, 1 + 3 = 4.
Root (276) = 2 + 7 + 6 = 15, 1 + 5 = 6.

The difference between the roots is 4–6 = –2, –2 + 9 = 7, while the root of the answer is 5 + 6 + 5 = 16, 1 + 6 = 7, that is, the check is successful.

Fast root calculation. When calculating the root, one can discard the numbers 9, as well as any groups of numbers that add up to 9.

Example. Calculate the root of number 8691. ~~8~~6~~91~~=6.

Exercises

Perform and verify the following multiplications:

45 × 45 = ?
95 × 95 = ?
205 × 205 = ?
33 × 37 = ?
104 × 106 = ?
104 × 107 = ? (take advantage of the fact that a×(b+1)=a×b+a)
81 × 89 = ?
82 × 89 = ?
Verify that the multiplication is correct: 345 × 291 = 100,395.

2. *Antyayor eva* (last two digits only)

This *sūtra* is dealing with multiplying by 11.

Consider the multiplication 17,453 × 11. The *sūtra* states that you need to take each digit of this number (from right to left), and add it with the previous one (to the right of it).

1) 17,453 attach the last digit to the answer: ...3.
2) 17,453 5 + 3 = 8, attach it to the answer: ...83.
3) 17,453 4 + 5 = 9, attach it to the answer: ...983.

4) 17,453 7 + 4 = 11, attach the last digit and remember the remaining one (1): ...1983.
5) 17,453 1 + 7 (+1) = 9, attach it to the answer: ...91,983.
6) 017,453 0 + 1 = 1, attach it to the answer: 191,983.

Root of the answer: 1 + 9 + 1 + 9 + 8 + 3 = 4.
Root of the first factor: 1 + 7 + 4 + 5 + 3 = 2.
Root of the second factor: 1 + 1 = 2.

2 × 2 = 4, which means that the verification is successful.

In the case when a two-digit number is multiplied by 11, this *sūtra* takes on an especially simple form: the digits of the multiplied number are written side-by-side and their sum is written between them:

23 × 11 = 253, 72 × 11 = 792.

If the sum of the digits of a two-digit number turns out to be greater than ten, then its last digit is written in the center, and the first one is added to the digit on the left:

59 × 11 = 649.

Exercises

Perform and verify the following multiplications:

3456 × 11 = ?
992 × 11 = ?
23 × 11 = ?
34 × 11 = ?
54 × 11 = ?
78 × 11 = ?

3. *Sopantyadvayamantyam* (double the penultimate digit plus the last)

Consider the multiplication 17,453 × 12. The *sūtra* states that one needs to take each digit of this number (from right to left), multiply it by two and add it with the previous one (to the right of it).

1) $17{,}45\underline{3}$ $3 \times 2 + 0 = 6$, attach it to the answer: ...6.
2) $17{,}4\underline{53}$ $5 \times 2 + 3 = 13$, attach the last digit and remember the remaining one (1): ...36.
3) $17{,}\underline{45}3$ $4 \times 2 + 5 + 1 = 14$, attach it to the answer: ...436.
4) $1\underline{7{,}4}53$ $7 \times 2 + 4 + 1 = 19$, attach it to the answer: ...9436.
5) $\underline{17}{,}453$ $1 \times 2 + 7 + 1 = 10$, attach it to the answer: ...09436.
6) $\underline{01}7{,}453$ $0 \times 2 + 1 + 1 = 2$, attach it to the answer: 209,436.

Root of the answer: $2 + 0 + 9 + 4 + 3 + 6 = 6$.
Root of the first factor: $1 + 7 + 4 + 5 + 3 = 11$; $1 + 1 = 2$.
Root of the second factor: $1 + 2 = 3$.

$2 \times 3 = 6$, which means that the verification is successful.

Exercises

Perform and verify the following multiplications:

$3456 \times 12 = ?$
$108 \times 12 = ?$
$78 \times 12 = ?$

4. *Ūrdhva Tiryagbhyām* (vertically and criss-cross)

This *sūtra* deals with the multiplication of arbitrary numbers.

Let's look at the multiplication of two-digit numbers first. The multiplication is performed according to the following scheme:

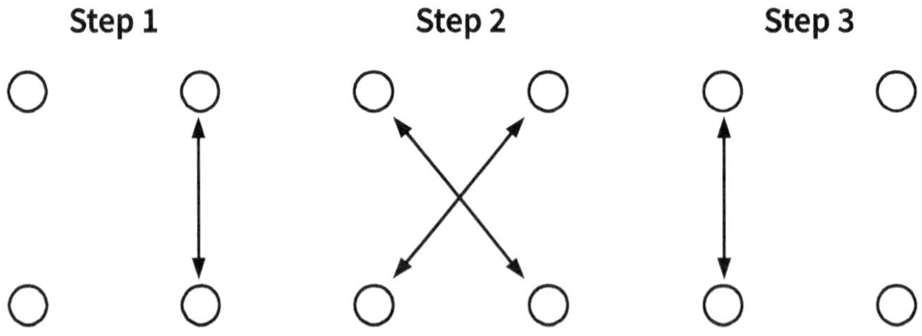

Figure 1. Multiplication of two-digit numbers.

Consider the application of this scheme using the example of 84×57:

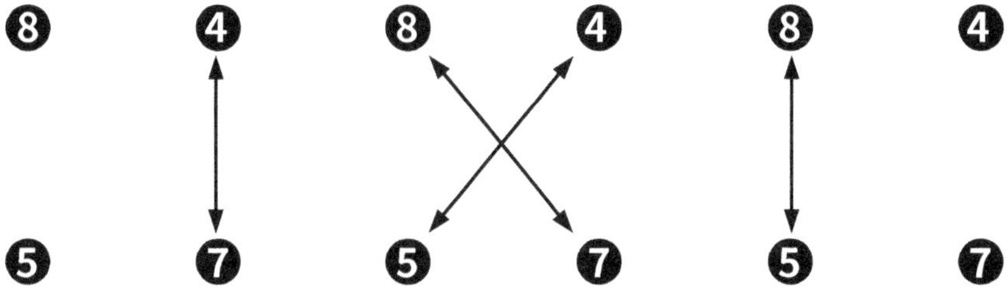

Figure 2. Multiplication of 84 and 57.

Step 1. $4 \times 7 = 28$. Attach the last digit and remember the remaining one (2): ...8.

Step 2. $8 \times 7 + 5 \times 4 + 2 = 56 + 20 + 2 = 78$. Attach the last digit and remember the remaining one (7): ...88.

Step 3. $8 \times 5 + 7 = 47$. Since this is the last step, both digits are attached: 4788.

Root of the answer: $4 + 7 + 8 + 8 = 27$, $2 + 7 = 9$.
Root of the first factor: $8 + 4 = 12$, $1 + 2 = 3$.
Root of the second factor: $5 + 7 = 12$; $1 + 2 = 3$.
$3 \times 3 = 9$, which means that the verification is successful.

Now consider the multiplication of three-digit numbers. The multiplication is performed according to the following scheme:

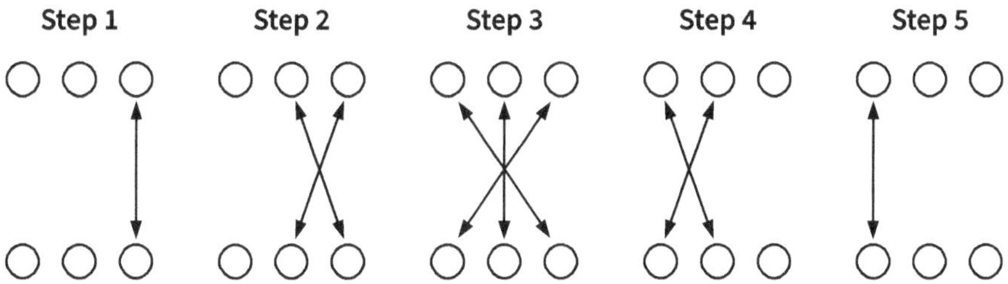

Figure 3. Multiplication of three-digit numbers.

Consider the application of this scheme using the example of 841×574:

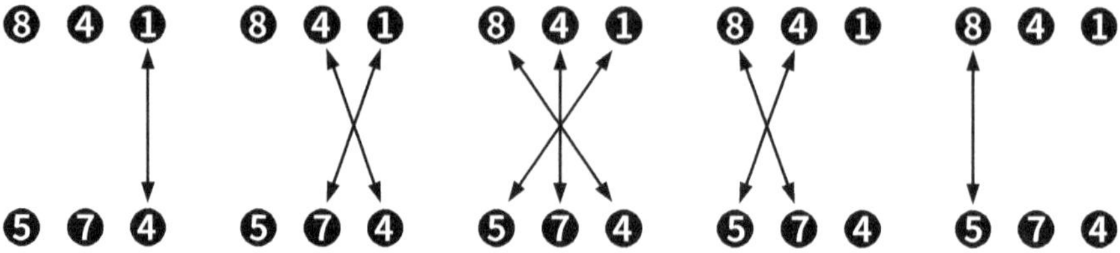

Figure 4. Multiplication of 841 and 574.

Step 1. $1 \times 4 = 4$. attach it to the answer: ...4.

Step 2. $4 \times 4 + 7 \times 1 = 16 + 7 = 23$. attach the last digit and remember the remaining one (2): ...34.

Step 3. $8 \times 4 + 4 \times 7 + 5 \times 1 + 2 = 32 + 28 + 5 + 2 = 67$. attach the last digit and remember the remaining one (6): ...734.

Step 4. $8 \times 7 + 5 \times 4 + 6 = 56 + 20 + 6 = 82$. Attach the last digit and remember the remaining one (8): ...2734.

Step 5. $8 \times 5 + 8 = 40 + 8 = 48$. Since this is the last step, both digits are attached: 482,734.

Root of the answer: $4 + 8 + 2 + 7 + 3 + 4 = 19$, $1 + 9 = 1$.
Root of the first factor: $8 + 4 + 1 = 13$, $1 + 3 = 4$.
Root of the second factor: $5 + 7 + 4 = 16$, $1 + 6 = 7$.
$4 \times 7 = 28$, $2 + 8 = 10$, $1 + 0 = 1$, which means that the verification is successful.

Now let us consider the graphic version of the *ūrdhva tiryagbhyāṃ sūtra*. This method is not tied to the number of digits in the multiplied numbers. Consider an example of multiplying a three-digit number by a two-digit number: 302×23.

Let's represent the first number with vertical lines, and the second with horizontal lines (we'll display zero with a dotted line):

Now let's start counting the intersections along the diagonals, bottom-right to top-left:

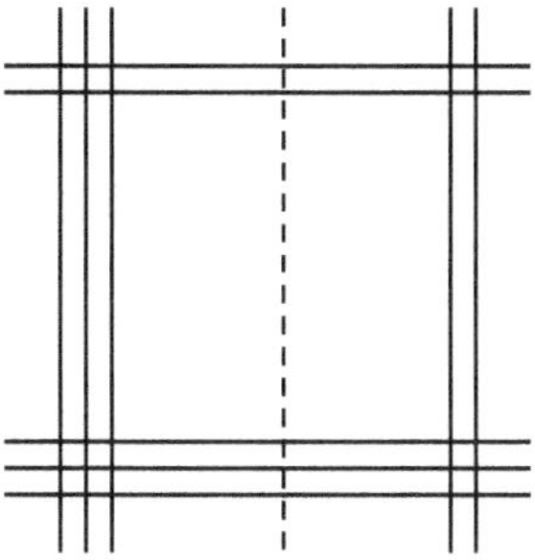

Figure 5. Multiplication of 302 and 23.

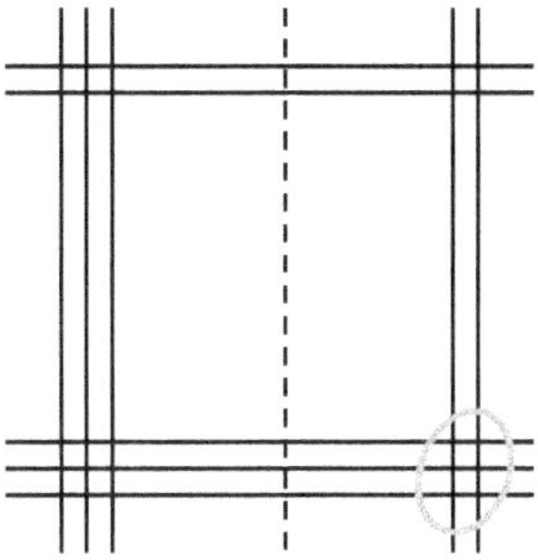

Figure 6. Intersection count is 6.

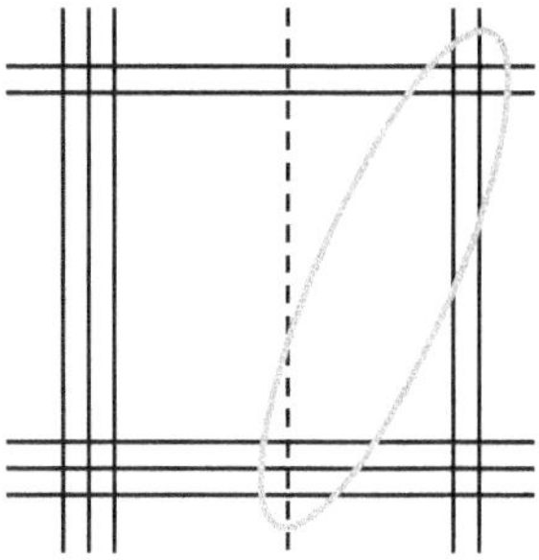

Figure 7. Intersection count is 4.

Step 1. Attach the number of intersections (6) to the answer: ...6:

Step 2. Attach the number of intersections (4) to the answer (ignoring intersections with the dotted line): ...46.

Step 3. Attach the number of intersections (9) to the answer (ignoring intersections with the dotted line): ...946.

Step 4. Attach the number of intersections (6) to the answer: 6946.

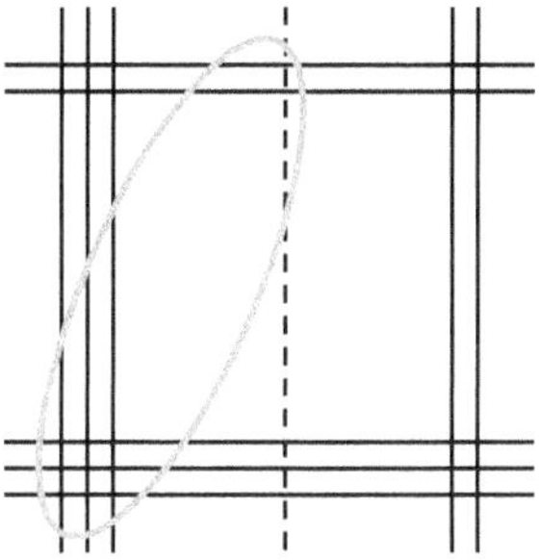

Figure 8. Intersection count is 9.

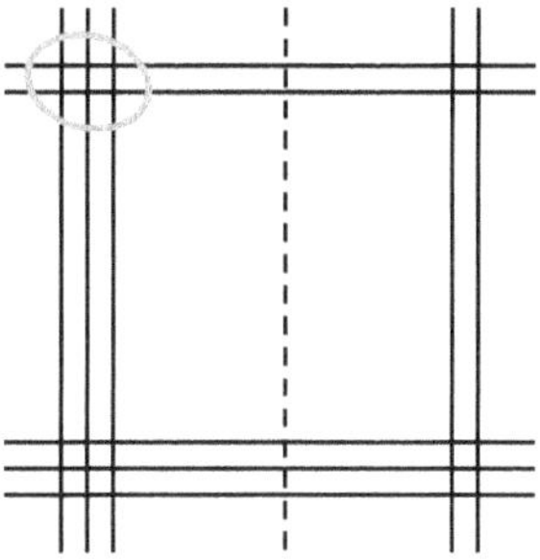

Figure 9. Intersection count is 6.

Root of the answer: $6 + 9 + 4 + 6 = 25$, $2 + 5 = 7$.

Root of the first factor: $3 + 0 + 2 = 5$.

Root of the second factor: $2 + 3 = 5$.

$5 \times 5 = 25$, $2 + 5 = 7$, which means that the verification is successful.

Exercises

Perform and verify the following multiplications. In accordance with the rule for the multiplication of two-digit numbers:

$46 \times 72 = ?$
$84 \times 23 = ?$

In accordance with the rule for the multiplication of three-digit numbers:

$781 \times 124 = ?$

In accordance with the graphical method:

$231 \times 112 = ?$
$104 \times 21 = ?$

5. Multiplication of numbers close to a round number (base)

In the case when the multiplied numbers are close to some round number called "base" (for example, 100, 1000, 10,000, etc.), the calculations can be significantly reduced.

Consider the multiplication 89×84.

Next to each number we write down its complement (with a "–" sign) to the nearest round number (100): Obviously, the complement of 89 is –11, and the complement of 84 is –16.

Figure 10. Multiplication of 89 and 84.

Step 1. Multiply the complements: $(-11) \times (-16) = 176$ (by the *antyayor eva sūtra*). Attach the last two digits to the answer: ..76, and remember the remaining one (1).

Step 2. Add the numbers along the first or the second diagonal (the result will be the same) and add the stored number (1): 89-16 (+1) = 74. Attach it to the final answer: 7476.

Root of the answer: $7 + 4 + 7 + 6 = 24$, $2 + 4 = 6$.
Root of the first factor: $8 + 9 = 17$, $1 + 7 = 8$.
Root of the second factor: $8 + 4 = 12$, $1 + 2 = 3$.
$8 \times 3 = 24$, $2 + 4 = 6$, which means that the verification is successful.

Now consider the product of a three-digit number and a four-digit number, one of which is greater than the base (1000), and the other is less than the base: 989×1012.

Next to each number, write its base's complement (1000): the complement of 989 is –11, and the complement of 1012 is 12.

Step 1. Multiply the complements: $(-11) \times 12 = -132$. Since the number turned out to be negative, attach to the answer its complement to the base:... 868, and remember –1.

Figure 11. Multiplication of 989 and 1012.

Step 2. Add the numbers along the first or the second diagonal (the result will be the same) and add the stored number (- 1): 989 + 12 + (- 1) = 1000. Attach it to the final answer: 1,000,868.

Root of the answer: $1 + 0 + 0 + 0 + 8 + 6 + 8 = 23$, $2 + 3 = 5$
Root of the first factor: $9 + 8 + 9 = 26$, $2 + 6 = 8$.
Root of the second factor: $1 + 0 + 1 + 2 = 4$.
$8 \times 4 = 32$, $3 + 2 = 5$, which means that the verification is successful.

5. Vedic Geometry

Vedic geometry is very practical in nature. The main surviving treatise on Vedic geometry is the *Śulbasūtras*, which describe the properties of geometric figures and give the rules for constructing altars.

The *Śulbasūtras* are part of *Kalpa*, one of the six *Vedāṅgas*. One of the meanings of the Sanskrit word *śulba* is "cord" or "rope." This is because measurements and construction of geometric figures were done by drawing arcs with different radii and centers using ropes.

Vedic geometry was fundamentally different from, for example, ancient Greek geometry in that it was not interested in obtaining "knowledge for the sake of knowledge itself." Ancient Greek scholars often disdained the practical aspects of mathematics. To prove this, it is enough to remember that one of the central questions of both ancient Greek and Vedic geometry was the construction of altars. This is where the classical problems of antiquity came from: squaring a circle,[10] trisecting an angle,[11] and doubling a cube.[12] At the same time, ancient Greek scholars rejected methods that could not be ideally implemented using compasses and a ruler, while the Vedic *Śulbasūtras* contain many effective practical techniques that allow the construction of complex geometric designs of altars. The *Śulbasūtras* begin with information on geometry and arithmetic, and end with a detailed description of the construction of altars from elementary geometric figures.

---

10 The task of squaring a circle was to find a way to construct a square equal in area to a given circle using a compass and a ruler. In 1882, the German scientist Karl Lindemann proved the transcendence of the number π and, as a consequence, the impossibility of an "ideal" solution to this problem with a compass and a ruler.

11 The problem of trisection of an angle was to divide an arbitrary angle into three parts using a compass and a ruler, and the impossibility of its solution was proved in 1837 by the French mathematician Wanzel.

12 Doubling a cube, also known as the Delian problem, requires the construction of the edge of a cube whose volume is double that of the given cube.

Construction of a square with a given side

Most of the geometric procedures described in the *Śulbasūtras* begin with the construction of the *prācī* – the segment of the altar symmetry line directed from east to west. The problem considered in this subsection means constructing a square with a vertical axis of symmetry in the form of a segment EW. Construction stages include:

1. Draw two circles with centers at points E and W and radius EW. The intersection points of these circles are denoted as X and Y. The midpoints of XP and PY segments are denoted as Z and T, respectively.
2. Draw four circles with centers at points Z, E, T, W and a radius of 0.5 EW. The intersection points of these circles form the required square ABCD.

This method of constructing a square is shown in the figure.

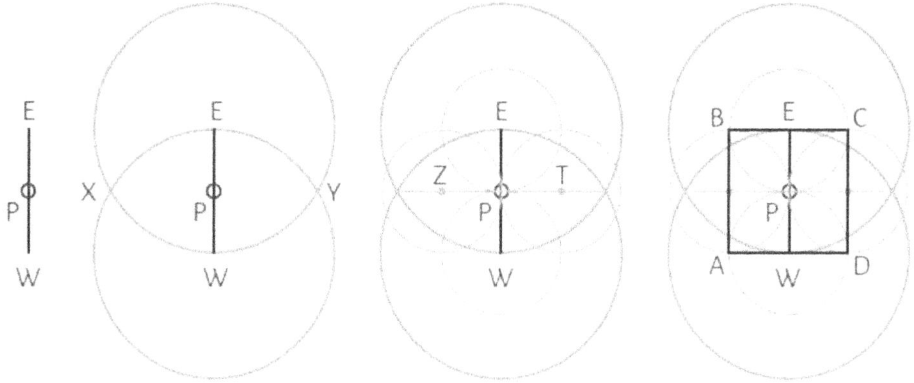

Figure 12. Stages of building a square with the axis of symmetry EW.

Theorem on the square of the diagonal ("Pythagorean" theorem)

The theorem about the square of the diagonal of a rectangle is the prototype of what later became known as the Pythagorean theorem and has the following formulation: "The area of a square built on the diagonal of a rectangle is equal to the sum of the areas of squares built on its sides." In the *Śulbasūtras* there are the examples of rectangles with integer sides:

$(3, 4, 5)\ 3^2 + 4^2 = 5^2$
$(12, 5, 13)\ 12^2 + 5^2 = 13^2$
$(24, 7, 25)\ 24^2 + 7^2 = 25^2$

Figure 13 is an illustration of this theorem. Let a right-angled triangle ABC be given. The area of a square built on its hypotenuse is equal to the sum of the areas of the ACFG and BIHC squares built on its sides.

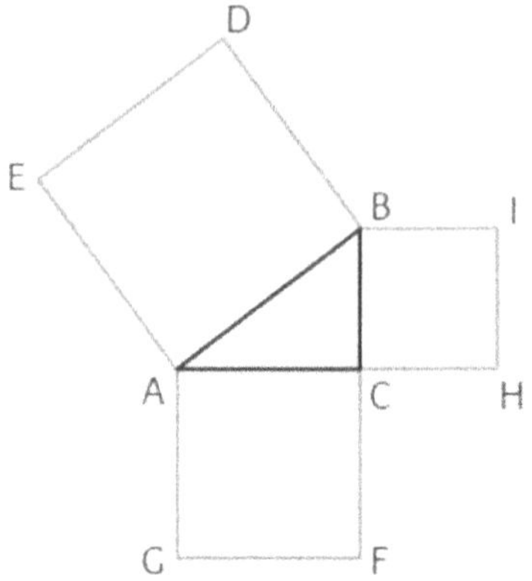

Figure 13. Prototype of "Pythagorean" theorem.

Construction of a square equal to the sum of two unequal squares

The solution to this problem is based on the previously considered theorem about the square of the diagonal of a rectangle. So, consider two squares: ABCD and EFGH. Put on the side BC of the larger of the squares the segment BJ equal to the side of the smaller square. It follows that $AJ^2=AK^2+KJ^2=AB^2+EF^2$, that is, AJ is a side of the required square.

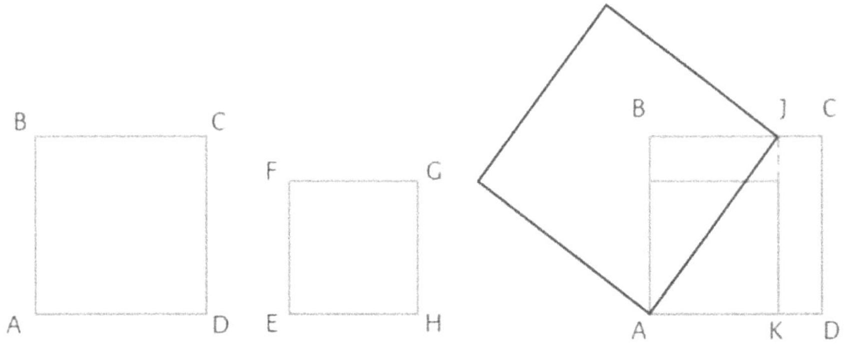

Figure 14. Sum of two squares.

Construction of square equal to the difference of two squares

The solution to this problem is also based on the previously considered theorem about the square of the diagonal. Let us construct a square

AIJK inside the square ABCD, which is a copy of the square EFGH. Let's draw an arc with center at point A and radius AB. The point of intersection of this arc with the side JK is denoted by L. It follows that

$$LK^2 = AL^2 - AK^2 = AB^2 - EF^2,$$

that is, LK is the side of the required square.

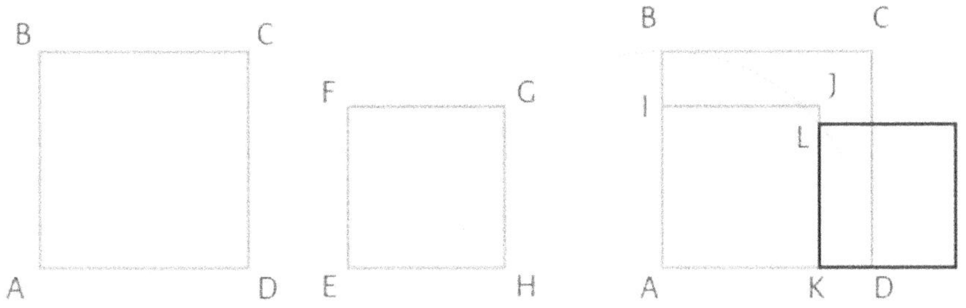

Figure 15. Difference of two squares.

Converting a rectangle into a square

Let the rectangle ABCD be given. The task is to construct a square with an area equal to the area of ABCD. The algorithm for solving this problem, described in the *Śulbasūtras*, has the following form:

1. Make the square ABEF (b).
2. Draw the segment GH, where the points G and H are the midpoints of the segments EC and FD, respectively (c).
3. Move rectangle HGCD to the top of square ABEF, forming rectangle BIJE (d).

Draw an arc centered at point F and radius FJ. The point of intersection of this arc with the side GH is denoted by K. Then the segment KH is the side of the required square (e).

To prove this, we denote the sides of the original rectangle ABCD as a and b $(a > b)$.

Then the square of the side KH can be represented as:

$$KH^2 = FK^2 - FH^2 = FJ^2 - FH^2 = (a+b)^2/2 - (a-b)^2/2 = ab,$$

that is, the area of the resulting square is equal to the area of the original rectangle ABCD, which was to be proved.

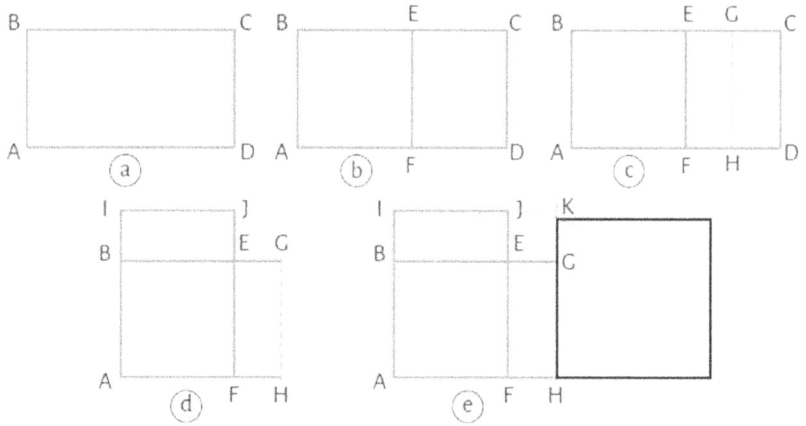

Figure 16. Rectangle to a square.

Converting a square into a circle

Let the square ABCD be given. The task is to build a circle with an area equal to the area of ABCD. Let's denote the center of the square as O. The algorithm for solving this problem, described in the *Śulbasūtras*, has the following form:

1. Draw an arc through points B and C with center at point O.
2. Draw radius OG perpendicular to side BC and intersecting it at point F. Select point E on it, dividing line segment FG in the ratio 1:2. Then OE is the radius of the desired circle.

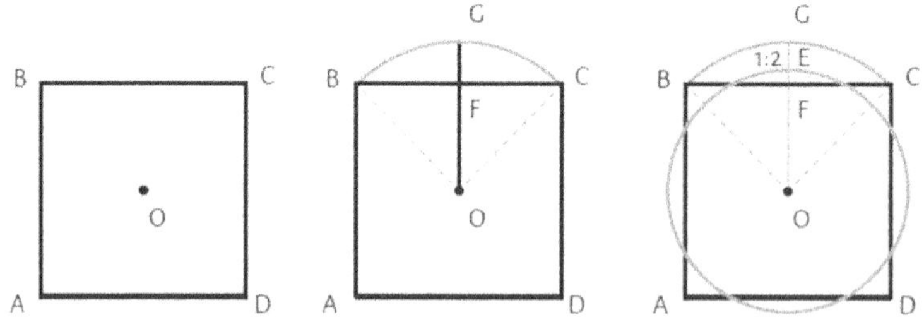

Figure 17. Square to a circle.

In order to prove this statement, we denote the side of the original square ABCD as a. Then the radius OE of the required circle is

$$r = \frac{a}{2} + \frac{1}{3} \times \frac{a\sqrt{2} - a}{2} = \frac{a(2 + \sqrt{2})}{6}$$

Accordingly, the area of the circle is

$$S = \pi r^2 = \pi a^2 \left(\frac{2 + \sqrt{2}}{6} \right)^2 \approx 1.017a^2,$$

i.e., differs from the area of the square by 1.7%, which is an acceptable error in practice.

Computation of the square root of two

The problem of approximately calculating the square root of two naturally arises, for example, when calculating the diagonal of a square with a unit side.

The rule established by the *Śulbasūtras* for calculating the diagonal of a square with unit side is as follows: "Increase the length [of the side] by its third and this third by its own fourth less the thirty-fourth part of that fourth." Literally, this means the following approximation to the root of two:

$$\sqrt{2} \approx 1 + \frac{1}{3} + \frac{1}{3 \times 4} - \frac{1}{3 \times 4 \times 34} = 1.4142156...,$$

which differs from the exact value by no more than 0.000002 (two millionths).

Note that this rule for calculating the square root of two is not a "random" approximation, but corresponds to the application of Newton's method[13] to solve the equation $x^2 = 2$ when choosing an initial approximation $x_0 = 4/3$.

13 The method of iterative solution of equations, applied by Newton at the end of the 17th century.

General information about the construction of altars

According to the *Śulbasūtras*, each altar is an aggregate of five levels. The odd (first, third, fifth) and even (second, fourth) levels have the same shape. To describe the procedures for their construction, the following units of length are used: *aṅgula* ("finger", 1.91 cm) and *puruṣa* (228.6 cm). One *puruṣa* consists of 120 *aṅgulas*. The height of each layer is 6.4 *aṅgulas*. The layers are designed in such a way that their inner joints do not touch. This requirement is difficult to fulfill, but adds strength to the structure and contributes to its aesthetics. Examples of *śyena* (eagle) and *rathacakra* (chariot wheel) altars are shown in the following figures.

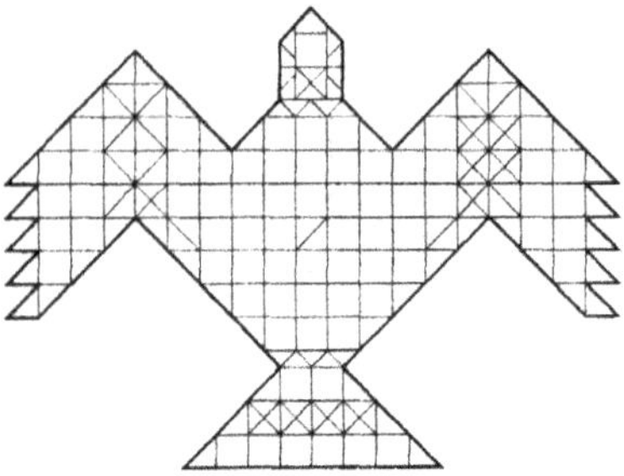

Layers 1, 3, 5

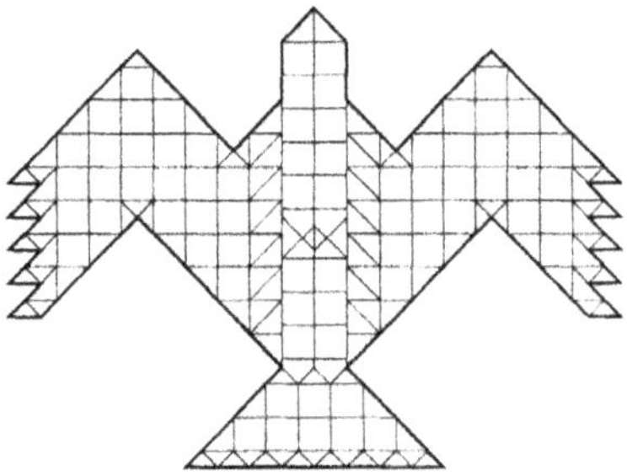

Layers 2, 4

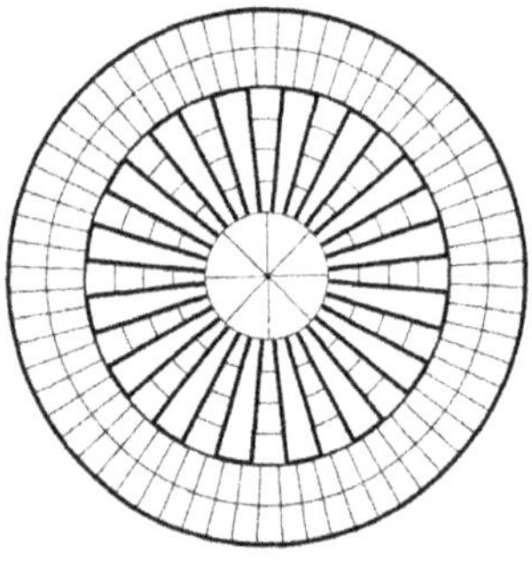

Layers 1, 3, 5

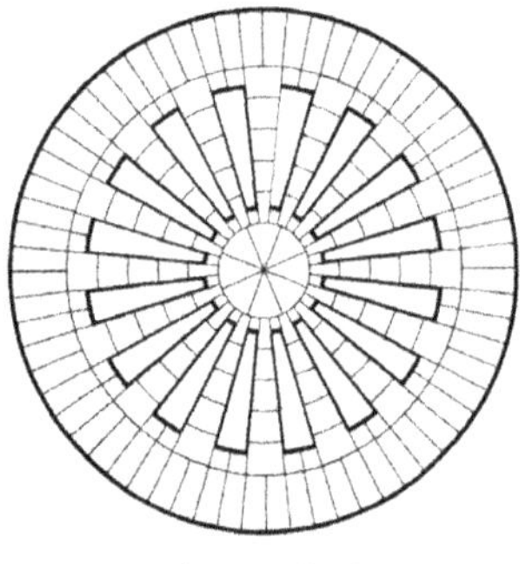

Layers 2, 4

Figure 18. Layers of *śyena* altar (top) and *rathacakra* altar (bottom).

6. Vedic Trigonometry

A significant amount of modern trigonometry is predated in the Vedic astronomical treatises (*Siddhāntas*[14]). Formerly, the sources for modern trigonometry were considered to be Arab and ancient Greek scholars. However, currently it is recognized that in the ancient Vedic text *Sūrya-siddhānta*, as well as in the texts of Āryabhaṭa and other Indian astronomers, we find such concepts as:

- division of a circle into 360 degrees,
- division of degrees into minutes
- sine and cosine tables
- basic trigonometric identities

In modern school programs, sines and cosines are determined by using the ratios of the sides of a right triangle or through the coordinates of the points on the unit circle. Similarly, in the *Sūrya-siddhānta*, the sine and cosine (*jyā* and *koṭi-jyā*) of an angle are defined as the vertical and horizontal coordinates of the points on a circle with a radius of 3438.

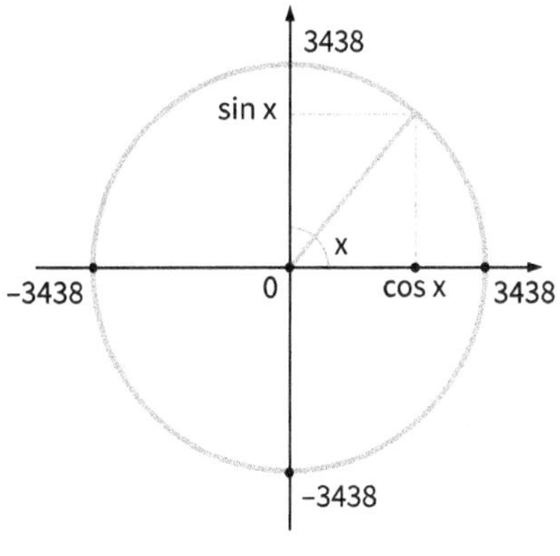

Figure 19. Sine and cosine in the *Sūrya-siddhānta*

This means that each sine is 3438 times larger than its modern counterpart. This also means that if the angles are given in arc minutes, then the sine of a small angle is almost equal to the measurement of that angle. This useful function is achieved in modern mathematics by measuring angles in radians, a technique first used in Europe by Roger Coates only in 1714.

Another feature of the number

14 The word *siddhānta* literally means "ultimate conclusion" or "perfect conclusion".

3438 is that it gives an approximation to the number π. Indeed, the number of minutes in a circle, 21,600 (i.e. 360 degrees × 60 minutes / degree), divided by 2π is 3437.746 or, rounded to the nearest integer, 3438. Thus, to perform trigonometric calculations, the number 3438 is a natural choice for the circle radius.

The table below contains the modern values and those given in the *Sūrya-siddhānta* (SS) for the sine function for angles with a step of 3.75 degrees. For comparison, the table also shows the current values, increased 3438 times. From this table it can be seen that the modern values practically coincide with data in the *Sūrya-siddhānta.*

TABLE 3

angle, degrees	angle, minutes	modern	modern × 3438	SS	difference
3.75	225	0.0654	225	225	0
7.5	450	0.1305	449	449	0
11.25	675	0.1951	671	671	0
15.0	900	0.2588	890	890	0
18.75	1125	0.3214	1105	1105	0
22.5	1350	0.3827	1316	1315	1
26.25	1575	0.4423	1521	1520	1
30.0	1800	0.5000	1719	1719	0
33.75	2025	0.5556	1910	1910	0
37.5	2250	0.6088	2093	2093	0
41.25	2475	0.6593	2267	2267	0
45.0	2700	0.7071	2431	2431	0
48.75	2925	0.7518	2585	2585	0
52.5	3150	0.7934	2728	2728	0
56.25	3375	0.8315	2859	2859	0
60.0	3600	0.8660	2977	2978	-1
63.75	3825	0.8969	3083	3084	-1
67.5	4050	0.9239	3176	3177	-1
71.25	4275	0.9469	3256	3256	0
75.0	4500	0.9659	3321	3321	0

angle, degrees	angle, minutes	modern	modern × 3438	SS	difference
78.75	4725	0.9808	3372	3372	0
82.5	4950	0.9914	3409	3409	0
86.25	5175	0.9979	3431	3431	0
90.0	5400	1.0000	3438	3438	0

Interestingly, the *Sūrya-siddhānta*, *sūtra* 2.16, contains a rule for recursively calculating the sines presented in this table: "To find an element of the sine table, take the sum of the previous and first elements of the table and subtract from it the sum of all previous ones divided by the first element."

Indeed, this rule can be proved using the trigonometric identity

$$\sin(n+1)x = \sin nx + \sin(x) - (\sin x + \sin 2x + \ldots + \sin nx) \times 2 \times (1-\cos x),$$

and also noting that

$$2 \times (1 - cos\,(3.75^o)) \approx \frac{1}{3438 sin(3.75^o)}.$$

So, for example, to get the fourth element of the sine table, we calculate:

$$671 + 225 - (671 + 449 + 225)\ /\ 225 = 890.02,$$

which is indeed a good approximation for the exact value (890).

Trigonometric series of Mādhava

Probably the most famous school of Indian mathematics is the *paramparā* (successive chain of teachers) dating back to Mādhava (c. 1340 – c. 1425). Geographically, these scientists lived in the South Indian territory called Kerala. The main representative of the Kerala school of mathematics, Mādhava, is famous for his contributions to a number of topics, but he is especially well-known for his trigonometric series, discovered in Europe only in the 17th century by Isaac Newton and James Gregory.

The most memorable of Mādhava's results is that the number π can be approximated by the infinite series:

$$\pi = 4 \times \left(1 - \frac{1}{3} + \frac{1}{5} - \frac{1}{7} + \frac{1}{9} - \frac{1}{11} + \ldots\right).$$

This formula allows you to calculate the number π with arbitrarily high accuracy by adding and subtracting rational numbers, although the number π itself is irrational and, moreover, transcendental (i. e., not the root of any integer polynomial).

Mādhava was also famous for encoding the coefficients of his formulas in poetic forms.

7. Vedic Combinatorics

As is known, combinatorics is a branch of mathematics that explores various ways to select a given number of elements from a certain finite set: placement, combination, permutation. Combinatorial methods are used in statistics, genetics, linguistics, information theory, and many other sciences.

The emergence and development of combinatorics in Europe in the 17th century coincided with a popular interest in gambling and attracted the attention of the famous mathematicians Pierre Fermat and Blaise Pascal. At the same time, numerous researchers, starting with Leonard Euler in the 18th century, showed that methods for solving combinatorial problems existed in ancient Indian mathematics.

One of the *ācāryas* (teachers) of Vedic combinatorics is Piṅgala, who lived in the 1st century BCE and is known by his work *Chandaḥśāstra* (*Chandaḥsūtra*), a treatise in Sanskrit on versification. This work is traditionally considered one of the *Vedāṅgas*.

According to Vedic phonetics, sounds are divided into two categories: long (*guru*) and short (*laghu*). Short sounds are traditionally denoted as ∪, and long sounds as —. For example, the word *Bhagavān* (the Supreme Lord) has a phonetic pattern ∪∪— because its last syllable is long and the first two are short. Similarly, the word *Śulbasūtra* has a phonetic pattern of —∪—∪, since its first and third syllables are long (according to the rules of *chandas*), and the second and fourth are short.

For example, the metric pattern of verse 2.13 of the *Bhagavad-gītā*:

dehino 'smin yathā dehe
kaumāraṁ yauvanaṁ jarā
tathā dehāntara-prāptir
dhīras tatra na muhyati

has the following form:

—∪——∪———

————∪—∪—

∪———∪———

———∪∪—∪—

Figure 20. Metric pattern for *Bg*. 2.13.

(This verse meter is called *anuṣṭup*.)

In his work, Piṅgala considered all possible combinations for a stanza of n syllables and wrote them down in the table of meters, shown in the figure for $n = 3$:

1. ∪∪∪
2. ∪∪—
3. ∪—∪
4. ∪——
5. —∪∪
6. —∪—
7. ——∪
8. ———

Figure 21. All combinations of meters for n=3

To construct such tables, Piṅgala used the rules corresponding to the modern rules for converting numbers to a binary number system from a decimal, and vice versa.

So, for example, a sequence

∪—∪——∪∪—

was interpreted as a binary number 01011001, which corresponds to the decimal number 89 ($89=0\times2^7+1\times2^6+0\times2^5+1\times2^4+1\times2^3+0\times2^2+0\times2^1+1\times2^0$) and has the same number (89) in the Piṅgala table for sizes of 8 syllables. Piṅgala also gave a method for quickly searching the phonetic pattern with a given number.

Piṅgala investigated the question of how many words of n syllables have a given number, k, short syllables and came up with ratios corresponding to modern formulas for combinations without repetitions:

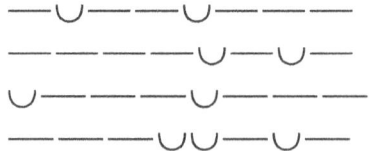

$$C_n^k = \frac{n!}{(n-k)!\,k!}.$$

So, for example, the number of phonetic patterns of 8 syllables that have 5 short syllables is equal to the number of combinations without repetitions from 8 to 5:

$$C_8^5 = \frac{8!}{(8-5)!\,5!} = \frac{6 \times 7 \times 8}{1 \times 2 \times 3} = 56,$$

where the symbol "!" means factorial – the product of all numbers not exceeding a given one (for example, $3! = 1 \times 2 \times 3 = 6$).

Piṅgala also found that the sum of all combinations without repetitions for a given number of n syllables equals 2^n (the modern name of this rule is Newton's binomial):

$$C_n^0 + C_n^1 + C_n^2 + \ldots + C_n^{n-1} + C_n^n = 2^n.$$

Or, considering that $C_n^0 = 1$:

$$C_n^1 + C_n^2 + \ldots + C_n^{n-1} + C_n^n = 2^n - 1.$$

The application of this rule is illustrated by the following problem:

> The Rajaḥ's palace had eight doors; these doors can either be opened one at a time, or two, or three etc. Question: what is the total number of all doors' opening options?

To solve this problem, one can directly add up the quantities of all combinations:

$$C_8^1 + C_8^2 + \ldots + C_8^7 + C_8^8 = 8 + 28 + 56 + 70 + 56 + 28 + 8 + 1 = 255$$

or immediately get an answer using the Pingala formula to count all:

$$2^8 - 1 = 256 - 1 = 255.$$

Another interesting example can be found in the work of the ancient Indian physician Suśruta (6th century BC). The *Suśruta-saṁhitā* states that there six main tastes. Further, the number of different tastes, taken by one, two, etc., from the six main tastes, is 63. As one can easily see, this is in full accordance with the above Piṅgala's formula, since $2^6 - 1 = 64 - 1 = 63$.

Piṅgala also describes *Meru-prastāra,* the prototype of the modern Pascal triangle, each level of which consists of all kinds of combinations without repetitions from a given number of elements (see the figure below). A wonderful property of this triangle is that each of its elements is the sum of those two above it.

1
1 1
1 2 1
1 3 3 1
1 4 6 4 1
1 5 10 10 5 1
1 6 15 20 15 6 1
1 7 21 35 35 21 7 1

Figure 22. ***Meru-prastāra*** **(Pascal's triangle)**

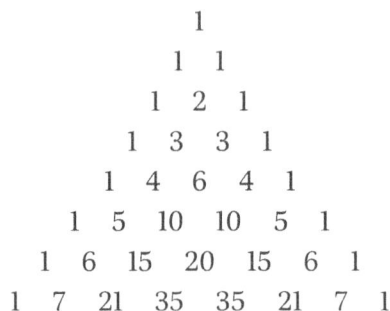

Figure 23. An example of Meru-prastāra, written in the Devanāgarī script. On the right are the sums of the elements of each level of the triangle: 1, 2, 4, 8, 16, 32, 64, 128, 256.

Another remarkable contribution of Piṅgala is the discovery and study of the sequence, now known as the Fibonacci numbers. Assuming that the duration of a short sound is 1 unit (one *mātrā*), and a long sound is 2 units, Piṅgala investigated the answer to the question: how many variants are there of sequences of single and double *mātrās* with a given total duration. Note that this problem also has its own musical analogy, where it is required to find the number of *tālas* (rhythms) with a given duration.

For example, for a total length of one unit, there is obviously only one variant: 1. For a length of two units, there are already two variants: 1 + 1 and 2. For three units there are 3 variants (1 + 1 + 1, 2 +1 and 1 + 2). For four units there are 5 variants (1 + 1 + 1 + 1, 2 + 1 + 1, 2 + 2, 1 + 2 + 1, 1 + 1 + 2). For five units, there are already 8 variants: 1 + 1 + 1 + 1 + 1, 1 + 2 + 1 + 1, 1 + 2 + 2, 1 + 1 + 2 + 1, 1 + 1 + 1 + 2, 2+ 1 + 1 + 1, 2 + 2 + 1, 2 + 1 + 2.

Continuing this sequence, Piṅgala established that each of its elements (starting from the third) is the sum of the two previous ones: 1, 2, 3, 5, 8, 13, 21, 34, 55, 89, 144, 233, 377, 610, 987, 1597, 2584, 4181, 6765, 10946, 17711, ...

Currently, this sequence bears its name after the Italian mathematician Leonardo of Pisa, Fibonacci, who is known to have studied Indian mathematical works in detail.

It is interesting that with the growth of this sequence, the ratio of its two neighboring elements tends to the famous golden ratio:

$$(\sqrt{5} + 1) / 2 = 1.61803...$$

The golden ratio is defined as the ratio of two values, in which the larger value relates to the smaller one in the same way as the sum of the values to the larger one. It can be verified that the ratio of two adjacent numbers in Pingala's sequence tends exactly to the golden ratio (2 : 1 = 2, 3 : 2 = 1.5, 5 : 3 = 1.67, 8 : 5 = 1.6, 13 : 8 = 1.625, 21 : 13 = 1.615 ...).

Magic squares

A magic square is a square table filled with different numbers in such a way that the sum of the numbers in each row, each column, and on both diagonals is the same. If in such a square the sums of numbers only in rows and columns are equal, then it is called semi-magic.

In Vedic culture, magic squares were considered as auspicious symbols of harmony. Images of these squares can still be seen inside and outside houses throughout India. The figure shows an example of such a square, made using the numbers of the Devanāgarī alphabet (for more details about the Vedic writing of numbers, see Chapter 3: "Decimal Numeral System").

Figure 24. Indian copper plaque with a 4x4 magic square

The simplest example of a 3 × 3 magic square is:

4	9	2
3	5	7
8	1	6

Figure 25.

As you can see, the sum of the numbers in each row, column, and also along the diagonals is 15.

Below is an example of a 4x4 magic square known from ancient Indian manuscripts.

1	14	15	4
12	7	6	9
8	11	10	5
13	2	3	16

Figure 26.

Also known is the magic square 4 × 4, reproduced on an engraving by Albrecht Durer:

Figure 27.

As one can see, the sum of the numbers in each row, column, and also along the diagonals of 4 × 4 magic squares is 34.

It is easy to show that the sum in each row or each column of an $n \times n$ magic square is:

$$S_n = \frac{n(n^2+1)}{2}.$$

So, for n = 5 this sum is 65, and for n = 10 it is 505.

Note that the problems of constructing magic squares have attracted and continue to attract the attention of mathematicians. Nārāyaṇa Paṇḍita, a representative of the mathematical school of Kerala, to which Mādhava also belonged (see Chapter 8: "Vedic Trigonometry"), was the first mathematician in history to single out the construction of magic squares as a separate discipline. In his work *Gaṇitakaumudī* ("Moonlight of Mathematics", or "Lotus of Mathematics") Nārāyaṇa Paṇḍita formulated the rules for constructing magic squares, and also used the combinatorial methods mentioned above in connection with the work of Piṅgala..

Interestingly, there are no universal methods for constructing magic squares of even order. One such particular method is shown below. To build a 4 × 4 magic square, one should:

A. Arrange the numbers from 1 to 16 in the cells in order.

B. Reverse the order of rows 2 and 4 and swap lines 2 and 3.

C. Reverse the order of the numbers in the second and third columns.

D. Reverse the order of numbers in rows 3 and 4.

A.

1	2	3	4
5	6	7	8
9	10	11	12
13	14	15	16

B.

1	2	3	4
12	11	10	9
5	6	7	8
16	15	14	13

C.

1	15	14	4
12	6	7	9
5	11	10	8
16	2	3	13

D.

1	15	14	4
12	6	7	9
8	10	11	5
13	3	2	16

Figure 28. Constructing a magic square

This method of constructing a 4 × 4 square is individual and does not apply to squares of other orders. However, for any square of odd order, there is a general and simple method for constructing them. Let's consider it using the example of constructing a 5 × 5 magic square.

1. Build a square with 25 cells and temporarily add it to a symmetrical stepped figure, as shown in the illustration.
2. In the resulting figure, place in sidelong rows "from top to bottom to right" 25 numbers from 1 to 25.
3. Each number outside the main square should be moved along the same row or column by the number of cells equal to the order of

the square (in this example, 5). For example, the number 6 should be placed in the cell under the number 18, the number 24 should be placed above the number 12, etc. As a result, the desired magic square is obtained, the sums of the numbers in rows, columns and diagonals of which are equal to 65.

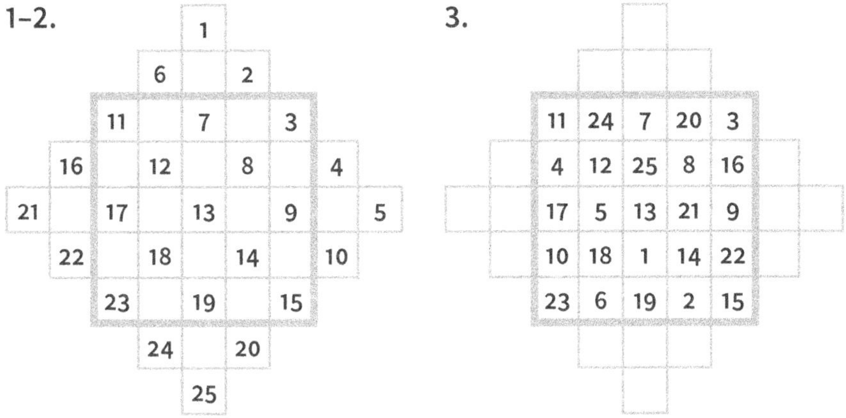

Figure 29. Method for constructing a 5x5 magic square.

8. On the Nature of Mathematical Creativity

Procreare iucundum sed partuire molestum.
("To beget is pleasant, but to give birth is painful.")
Carl Friedrich Gauss (1777–1855)

Practically everyone – scientists, artists, representatives of all professions in general – have thought about the nature of creativity, about the true source of ideas, inspiration, intuition, and insight. Understanding this issue is closely related to understanding the nature of personality and its connection with the outside world. In this context, the views of Vedic and modern science are quite different from each other. For example, the famous evolutionary biologist John Maynard Smith said that "the individual is simply a device constructed by the genes to ensure the production of more genes like themselves."[15] Prominent English inventor Alan Turing believed that "since whatever a human being can do a computer can imitate, a human being is merely a machine."[16] However, a civilized person is unlikely to be satisfied with the mechanistic explanation of the creative process as a set of biochemical reactions or electrical interactions. So what exactly *is* the nature of creative inspiration? Could seeking an answer to this question truly change our life?

A number of prominent scientists have studied the enigma of creative inspiration, such as Plato, Descartes, Spinoza, Popper, and Poincaré. Both René Descartes and Karl Popper have suggested that the processes taking place in consciousness are not limited by the brain. Trying to explain the process of creativity, the famous French mathematician and physicist Henri Poincaré proposed dividing the process into two stages: conscious and subconscious activity. In the book *Science and Method,* Poincaré expressed his idea of the "two selves" – the conscious self and the subliminal self. The subconscious self generates

15 Smith, J.M. (1988). "The Limitations of Evolution Theory." In: *Did Darwin Get It Right?*. Springer, Boston, MA. https://doi.org/10.1007/978-1-4684-7862-4_20

16 Turing, A. "Computing Machinery and Intelligence," *Mind*, 1950

many ideas, while selecting from them only the most fruitful, which then appear for the conscious self as an idea or inspiration.

Nevertheless, Poincaré, one of the most prominent scientists of his time, was not able to explain the nature of the subconscious self. Despite this inability, Poincaré praised the subliminal self as "possessing a sense of tact and beauty of a higher order, lying beyond sensory perception, understanding the harmony of the parts and their happy balance."[17]

Poincaré proposed that the subconscious self randomly iterates through many combinations of mathematical symbols until it finally finds a combination that meets the desire of the conscious self. However, such a mechanistic approach is not necessarily viable. In his book *Mechanistic and Nonmechanistic Science*, Richard Thompson shows that even if one assumes that one combination is selected in every cubic angstrom[18] of the brain every billionth of a second, the proof of a mathematical theorem of little complexity would take, according to minimal estimates, up to 100 years.

At the same time, at the moment of inspiration, the solution seems to appear instantaneously, rather than as the end result of working through colossal numbers of options. So, the phenomenon of inspiration cannot be easily explained merely on the basis of mechanistic models of brain activity that are consistent with modern theories of physics and chemistry. In this case, researchers tend to concentrate solely on the external side of the creative process, while overlooking more subtle processes lying outside the sphere of interest for most modern scientists.

In his book *Substance and Shadow: The Vedic Method of Knowledge*, Suhotra Swami concludes that scientific theories and discoveries often arise not so much from logical inferences as from a disordered, bizarre, or even mystical state of mind. Even empiricist philosophers sometimes admit this. Francis Crick, the eminent English molecular biologist, biophysicist, and neuroscientist, explains:

> It is not easy to convey, unless one has experienced it, the dramatic feeling of sudden enlightenment that floods the mind

17 Poincare, A. Science and Hypothesis, 1905. Walter Scott Publishing.

18 The angstrom is a metric unit of length equal to 10–10 m.

> when the right idea finally clicks into place. One immediately sees how many previously puzzling facts are neatly explained by the new hypothesis. One could kick oneself for not having the idea earlier, it now seems so obvious. Yet before, everything was in a fog.[19]

Indeed, if the phenomenon of inspiration is caused by the action of a neural algorithm, then why does it usually come as a sudden vision of a completed solution to a problem without awareness of the intermediate stages? The outstanding Indian mathematician Srinivasa Ramanujan, as a result of what could be called "religious revelations," formulated the most complex theorems, while other scientists many years later were able to confirm their results through many intermediate stages. It is known that Gottfried Hardy, a famous English mathematician, commented on the results reported to him by Ramanujan: "They must be true because, if they were not true, no one would have had the imagination to invent them."[20] The example of Mozart is also instructive: "[He] once described how he created his musical works:

> When I feel well and in good humor, or when I am taking a drive or walking, . . . thoughts crowd into my mind as easily as you could wish. Whence and how do they come? I do not know and I have nothing to do with it."[21]

An interesting claim by Richard Wagner is that it took his teacher some time to teach how to transform into music the images that suddenly appeared in his mind. The famous Danish physicist Niels Bohr said that the model of an atomic nucleus surrounded by orbitals of electrons came to him intuitively. Note that even Plato believed that creative ideas, *eidos*, come from the world of "Forms." One of the most famous astrophysicists of our time, Roger Penrose, adopts a similar concept in his book *The Road to Reality: A Complete Guide to the Laws of the*

---

19 Francis Crick, *What Mad Pursuit: A Personal View of Scientific Discovery* (1990) BasicBooks, NYC

20 E.H. Neville, "Srinivasa Ramanujan," *Nature*, Vol. 149, March 14, 1942.

21 Thompson, Richard L., *Mechanistic and Nonmechanistic Science* (LA: Bhaktivedanta Book Trust, 1981), 171.

Universe, and believes that "consciousness cannot be understood in terms of present day science."[22]

The appearance of ideas in the intellect as impressions correlates well with the concept of subtle sound, *paśyanti,* in the Vedic hierarchy of sound levels, which is then reflected on the level of the mind (*madhyama*) in the form of thought (*Śrīmad-Bhāgavatam,* 11.21.36). The outstanding American inventor and futurist, Nikola Tesla, said that he often experienced revelations, which he explained as his connection to the intellectual core of the universe. How can we develop a model to help explain the phenomenon of creative inspiration that incorporates the various concepts of these first-rate thinkers?

The philosophy of the *Bhagavad-gītā* offers a simple and logical explanation of the phenomenon of inspiration from the point of view of the law of *karma.* This is how Bhaktivedanta Swami Prabhupāda put it in a lecture on January 19, 1975:

> They are so many people are working for inventing something, but one man takes the credit; other man cannot. Why? Why this discrimination? There must be some cause of this discrimination. That is explained in the *Bhagavad-gītā, sarvasya cāhaṁ hṛdi sanniviṣṭaḥ*: "I am situated in everyone's heart." *Mattaḥ smṛtir jñānam apohanaṁ ca* (Bg. 15.15). *Mattaḥ,* "From Me, if I give you the intelligence, 'Now mix this chemical with this chemical, your product will come.'" Others, they have got the chemical, and the laboratory man also there. But one takes the credit; one cannot take the credit. That is due to Kṛṣṇa. If Kṛṣṇa gives the intelligence, then he gets the intelligence and he takes the credit.
>
> —*Śrīmad-Bhāgavatam* Lecture, 3.26.44, Jan. 19, 1975, Bombay

Bhagavad-gītā explains that such inequality is the result of *karma,* that is, the accumulated consequences of previous actions. The controlling aspect of the Absolute Truth, the Supersoul, takes into account not only the desires of the living entity, but also the results of his past activities. Paramātmā, the Supersoul, directs nature according to a higher code of laws. These laws are as real as the so-called "laws of nature" determined

22 R. Penrose, *The Road to Reality*, 2007, Vintage Books, New York.

by physics and chemistry, but they are much more complex than the latter and are directly related to living beings. As explained in the fifth chapter of *Bhagavad-gītā*, if the desire of the living being is sanctioned by Paramātmā, then its realization appears to be automatically fulfilled by the gross and subtle mechanisms of material nature.

Thus, we see that although modern mechanistic approaches have difficulty trying to account for the processes of inspiration and creativity, the philosophy of *Bhagavad-gītā* is able to provide a relatively simple and logical explanation for these phenomena without violating empirical laws.

9. Brief Information about Mathematicians of the Vedic Tradition

Āryabhaṭa (476–550)

Prominent astronomer and mathematician during the Gupta dynasty, the golden age of Indian history. His fundamental work, *Āryabhaṭīya*, which he wrote at the age of 23 in the year 3600 Kali-yuga (499 CE), has survived to our time. His main achievement is considered to be the systematization of knowledge in astronomy and mathematics as preserved by oral tradition. Āryabhaṭa himself wrote that all knowledge came to him by the grace of Svayambhū (Lord Brahmā). Mathematical knowledge contained in *Āryabhaṭīya* includes: the method of extracting the square and cube root in the decimal notation system, formulas for the area of a circle and the volume of a sphere, an approximate value for the number π: 3.1416, a rule for verifying the result of calculations using a root (see Chapter 4: "Vedic Arithmetic"), analogs of the geometric theorems of Pythagoras and Thales, formulas for solving quadratic equations, and rules for summing power series.

Brahmagupta (c. 598 – c. 668)

The main work of Brahmagupta is *Brāhma-sphuṭa-siddhānta* (*Improved Treatise of Brahmā*). As customary in the Vedic tradition, this work is written in verse and without proofs. This work examines solutions to Diophantine equations, gives an algebraic formula now known as the Brahmagupta-Fibonacci identity, and provides algorithms for multiplying numbers closely resembling multiplication methods taught in modern school programs. In the geometric part of his treatise, he gives an analog of Heron's formula for calculating the area of a triangle, rules for constructing triangles with rational sides, as well as the famous Brahmagupta theorem:

> If a cyclic quadrilateral has perpendicular diagonals intersecting at point M, then the line passing through point M and perpendicular to one of its sides, divides the opposite side in half.

It is also interesting that for an approximate calculation of certain functions, Brahmagupta used an analog of the Newton-Stirling formula from modern differential calculus.

Bhāskarācārya (1114–1185)

The most famous work of Bhāskarācārya is *Siddhānta-śiromaṇi* (*Crown of Treatises*), which consists of four parts: *Līlāvatī* (arithmetic), *Bījagaṇita* (algebra or root calculus), *Gōlādhyāya* and *Grahagaṇita* (the theory of planetary motion). Bhāskarācārya described computational rules, accompanied by many practical examples. He systematized methods for solving various equations, formulated preliminary concepts of modern mathematical analysis (integral and differential calculus) and an analog of the modern mean value theorem (Rolle's theorem). Also, Bhāskarācārya developed the basics of spherical trigonometry.

Piṅgala (c. 150 BC)

He is considered the founder of combinatorics and the first scientist to use the binary number system. More details about his work can be found in Chapter 7: "Vedic Combinatorics".

Mādhava (1350–1425)

Founder of the Kerala School of Astronomy and Mathematics. He became famous for his trigonometric series (see Chapter 6: "Vedic Trigonometry").

Śrinivasa Ramanujan (1887–1920)

An outstanding mathematician of the twentieth century. As a result of what he called "religious revelations," he obtained fundamental results in various fields of mathematics, especially in number theory. In 1913, he started a correspondence with the famous English mathematician Godfrey Hardy, as a result of which Hardy accumulated about 120 formulas unknown to science at that time. By Hardy's insistence, Ramanujan came to the University of Cambridge. Hardy is known to have commented on the results reported to him by Ramanujan: "They must be true because, if they were not true, no one would have had the imagination to invent them." Currently, the formulas derived by Ramanujan appear in modern branches of science that were unknown during his time. Ramanujan was elected a member of the Royal Society (English Academy of Sciences) and at the same time offered a professorship at the University of Cambridge, becoming the first Indian to be awarded such honors.

Bhaktisiddhānta Sarasvatī (1874–1937)

Astronomer and mathematician who later became a Vaiṣṇava guru, preacher, and spiritual teacher of the founder of the International Society for Krishna Consciousness (ISKCON) His Divine Grace Bhaktivedanta Swami Prabhupāda (1896–1977). In 1888, at the age of 14, he began translating from Sanskrit into Bengali, with personal commentary, on two classical astronomical treatises: *Siddhānta-śiromaṇi* (completed in 1893) and *Sūrya-siddhānta* (completed in 1896). Recognizing

his exceptional contributions to both Eastern and Western astrology and astronomy, he was awarded the title of Śrī Siddhānta Sarasvatī at the age of 15. In 1896 he founded and became the editor of several astronomical publications: the journals *Bṛhaspati* (*Scientific Indian*) and the *Jyotirvidyā*, the almanac *Bhakti-bhāvana Pañjikā*, and the Gauḍīya Vaiṣṇava calendar *Śrī Navadvīpa Pañjikā*.

Appendix. Vedic Units of Measure for Distance and Time

TABLE 4. Vedic units of distance and their modern equivalents*

Vedic unit of distance	Relation	Modern equivalent, m
paramanu		1×10^{-7}
anu	2 × *paramanu*	1.9×10^{-7}
trasarenu	3 × *anu*	5.8×10^{-7}
balagra (upper point of a hair)	8 × *trasarenu*	4.7×10^{-6}
likhya	8 ×*balagra*	3.7×10^{-5}
yuka	8 × *likhya*	0.0003
yava	8 × *yuka*	0.0024
angula (finger thickness)	8 × *yava*	0.019
pada (foot width)	6 × *angula*	0.11
pradeśa	11 × *angula*	0.2
vitasti (span between the extended thumb and the little finger)	2 × *pada*	0.23
hasta (elbow)	2 × *vitasti*	0.45
dhanu (bow)	4 × *hasta*	1.82
puruṣa	120 × *angula*	2.29
gavyūti	2000 × *dhanu*	3658
yojana		12880

* *Matsya Puranam*, Oriental Publishers, Delhi, 1972. p. 303.

TABLE 5. Vedic units of time and their modern equivalents

Vedic unit of time	Relation	Modern equivalent, sec.*
truṭi		8/13500
vedha	100 × *truṭi*	8/135
lava	3 × *vedha*	8/45
nimeṣa	3 × *lava*	8/15
kṣaṇa	3 × *nimeṣa*	8/5
kāṣṭhā	5 × *kṣaṇa*	8
laghu	15 × *kāṣṭhā*	120
daṇḍa	15 × *laghu*	1800 (30 min.)
muhūrta	2 × *daṇḍa*	3600 (60 min.)

* Multiply values in the third column by 0.8 to get equivalent values for the 48-minute *muhūrta*, which is also used in the *Bhāgavata* and other *Purāṇas*.

TABLE 6. Vedic units for measuring the duration of eras

Vedic unit of time	Relation	Modern equivalent (years)
divya-yuga		4,320,000
manvantara	71 × *divya-yuga*	306,720,000
kalpa (day of Brahmā)	14 × *manvantara*	4 320,000,000
parārdha (half of Brahmā's life)	50 × 360 × 2 × *kalpa*	155,520,000,000,000
lifespan of the universe	2 × *parārdha*	311,040,000,000,000

PART TWO

Nature of Time

Introduction

Time is one of the major mysteries of the universe. Scientists, philosophers, and theologians have always been fascinated by the nature of time and have been attempting to analyze it for centuries. And yet the goal remains elusive. This is not surprising. Eastern wisdom, as exemplified by the *Śrīmad-Bhāgavatam* – the quintessence of ancient Vedic literature – describes time as an inconceivable energy of the Supreme Absolute Truth.

We can see that the modern age of rapid scientific and technological progress in certain areas has not produced a holistic understanding of this basic foundation of the universe. What is time? Passive environment or driving force behind all processes of creation, maintenance, and destruction in the universe? The *Śrīmad-Bhāgavatam* gives a comprehensive answer to this question, in some ways confirming and in some ways refuting the statements of Hawking, Penrose, Einstein, St. Augustine, and other prominent thinkers in Western culture.

Unlike the rapidly shifting and apparently disparate knowledge offered by modern science, *Śrīmad-Bhāgavatam* presents an integral picture of the surrounding reality. In particular, the processes of creation and the functional principles of the material world are considered in the Third Canto of this work, in the dialogues of Śukadeva Gosvāmī and Mahārāja Parīkṣit, Maitreya and Vidura, as well as Lord Kapila and Devahūti. These conversations shed light on the nature of time as one of inconceivable energies of the Supreme Personality of Godhead. In the philosophy of the *Śrīmad-Bhāgavatam*, time is an integral part of a comprehensive worldview of the universe, which also includes the Supreme Controller (Paramātmā), living beings (*jīvas*), material nature (*prakṛti*), as well as activities (*karma*) and time (*kāla*).

According to the opening aphorism of the *Vedānta-sūtra* – *athāto brahma jijñāsā* – the driving force behind the narrative of *Śrīmad-Bhāgavatam* is the questions asked by the sages 5000 years ago during a thousand-year sacrifice being performed at the Naimiṣāraṇya forest. In *Bhāgavatam* verses 2.8.12–13, Mahārāja Parīkṣit asks Śukadeva Gosvāmī:

yāvān kalpo vikalpo vā
yathā kālo 'numīyate
bhūta-bhavya-bhavac-chabda
āyur-mānaṁ ca yat sataḥ

kālasyānugatir yā tu
lakṣyate 'ṇvī bṛhaty api
yāvatyaḥ karma-gatayo
yādṛśīr dvija-sattama

"Please explain the duration of time between creation and annihilation, and that of other subsidiary creations, as well as the nature of time, indicated by the sound of past, present and future. Also, please explain the duration and measurement of life of the different living beings known as the demigods, the human beings, etc., in different planets of the universe. O purest of the *brāhmaṇas*, please also explain the cause of the different durations of time, both short and long, as well as the beginning of time, following the course of action."

Further, in verse 3.10.10 Vidura asks Maitreya:

yathāttha bahu-rūpasya
harer adbhuta-karmaṇaḥ
kālākhyaṁ lakṣaṇaṁ brahman
yathā varṇaya naḥ prabho

"O greatly learned sage, kindly describe eternal time, which is another form of the Supreme Lord, the wonderful actor. What are the symptoms of that eternal time? Please describe them to us in detail."

Devahūti asks her son, Lord Kapila in verse 3.29.4:

kālasyeśvara-rūpasya
pareṣāṁ ca parasya te
svarūpaṁ bata kurvanti
yad-dhetoḥ kuśalaṁ janāḥ

"Please also describe eternal time, which is a representation of Your form and by whose influence people in general engage in the performance of pious activities."

This essay presents the point of view of the narrators of the *Śrīmad-Bhāgavatam*, along with similarities and differences in comparison to the views of modern science.

1. A Brief Historical Review

"The science of the twentieth century is at a stage when the moment has come for the study of time in the same way that matter and energy filling space are studied."
V. I. Vernadsky (*1863–1945*)

Saint Augustine is credited with the saying: "What is time then? If nobody asks me, I know; but if I were desirous to explain it to one that should ask me, plainly I do not know." This paradoxical statement may sound amusing, and yet perhaps it reflects a deeper reality – according to *Śrīmad-Bhāgavatam,* time is an incomprehensible aspect of the Absolute Truth. But Augustine also made a more specific and intriguing statement: time is an inalienable property of the universe created by God and therefore time did not exist before the emergence of the universe. This agrees with the statements of the leading figures of modern physics, Roger Penrose and Stephen Hawking, who argue that the known laws of the universe do not work at the moment of the Big Bang.[1] Therefore, it is useless to speculate about what happened before the moment of the Big Bang.[2]

Discussions about the nature of time have always been important to the development of science. Historically, two sorts of approaches have emerged:

1. relational – time as an abstraction for describing relationships between objects.
2. substantial – time as a substance independent of objects.

1 Big Bang is a cosmological model that describes the early development of the universe, namely, the beginning of its expansion, before which the universe was in a singular state, i.e., had an infinitely small size plus an infinitely high density and temperature. In 1949, the famous astrophysicist Fred Hoyle ironically called the model "Big Bang," after which the name was historically entrenched in the cosmology discipline.

2 Subsequently, Penrose changed his point of view, which is reflected in his work, *Cycles of Time: An Extraordinary New View of the Universe.*

Prominent representatives of the relational approach include Aristotle, Leibniz, and Einstein. For instance, Albert Einstein believed that "time is a convenient abstraction for describing the laws of motion."

One of the prominent representatives of the substantial approach was Isaac Newton (1643–1727), who argued in his famous work *Mathematical Principles of Natural Philosophy* that "absolute time, without any relation to anything external, flows uniformly."

At the beginning of the 20th century, the relational approach became dominant when Albert Einstein introduced within the framework of the theory of relativity[3] the concept of four-dimensional space-time, making time a coordinate that complements the three-dimensional space we perceive. Thus, just as the concept of ether was rejected as a result of the Michelson-Morley experiment, time was assigned the quality of an abstraction that does not have an independent existence from material objects. The main motive for combining time with three spatial dimensions was that the observed rate at which time passes depends on the object's velocity relative to the observer, as well as on the strength of the gravitational field, which can slow down the flow of time.

The more mass placed in a certain area of space, the more space-time is curved and the slower a nearby clock will run (from the point of view of an outside observer). At a certain critical mass, space-time curves so much that even light cannot escape its gravitational pull, and a black hole is created.

So, in space-time, each event is characterized by four coordinates – three spatial coordinates plus time. Thus, the coordinates determine both where and *when* events occur. According to organic chemist and spiritual leader Dr. Thoudam Damodara Singh (Bhaktisvarūpa Dāmodara Goswami), evolution is the movement of living beings along the space-time continuum.[4]

The movement of objects in space-time can be thought of as a generalization of movement in two-dimensional or three-dimensional space. So, the position of a point fixed in space will represent a straight line

---

3 The special and general theories of relativity, which Einstein developed in 1905 and 1915, respectively. These theories generalized the classical (nonrelativistic) Newtonian mechanics, while revising the role of time as a physical concept.

4 Bhaktisvarūpa Dāmodara Goswami, *Life and Origin of the Universe* (Kolkata, India: Bhaktivedanta Institute, 2004).

in space-time, and the circular motion of the sun around the earth will represent a helix in space-time, as shown in Figure 1.

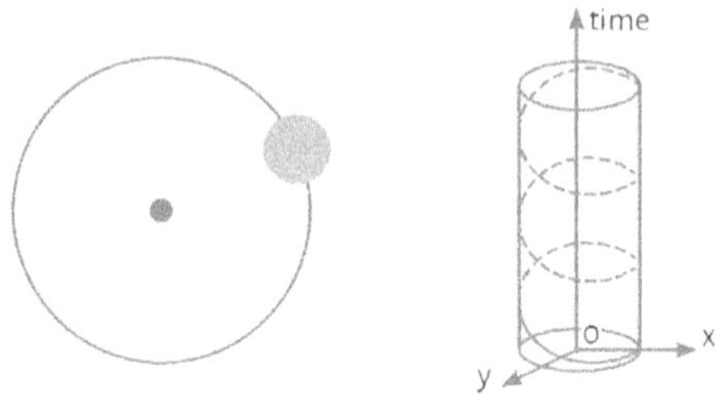

Figure 1

Unlike ordinary spatial coordinates, the concept of a light cone (see Figure 2) arises in space-time to impose restrictions on the permissible coordinates of events. The very fact of this limitation is caused by another postulate of the theory of relativity – the speed of light is constant and does not depend on the speed of its source. The speed of light equals 186,282 mi (299,792 km) /sec,[5] and according to theory, no object in the world can move at a higher speed.[6]

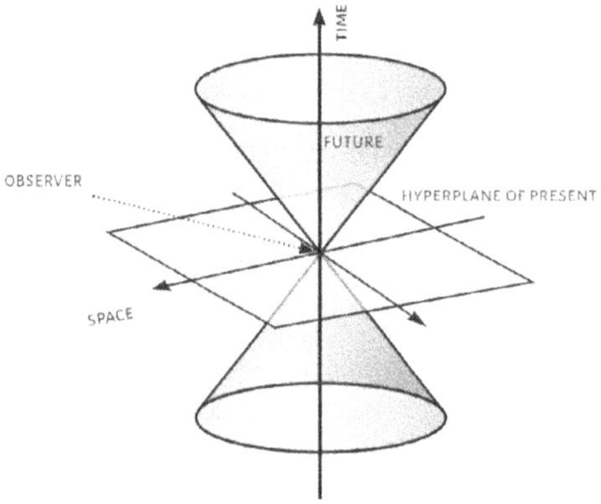

Figure 2: Observer, space, future, and hyperplane of present.

To visualize a light cone, assume we are in two-dimensional space, and time is a vertical coordinate. Let the observer be at the apex of the cone through which the horizontal plane of the present passes. Then:

1. Part of the surface in the region of the future in relation to the vertex contains all events that can be reached by a light signal from the apex.
2. The lower part contains all events in the past, such that the light signal emitted from them can reach the apex.

5 Although the modern value for the speed of light was obtained only at the beginning of the 20th century, in the commentary of Sāyaṇa (who lived in the 14th century in South India) to the *Ṛg Veda*, a value corresponding to 305,180 km/s was given, which differs by less than 2% from modern value.

6 Particle physics speaks of the presence of special particles, tachyons, moving with a velocity exceeding the speed of light. Also, the modern theory of inflationary expansion of the universe posits that the expansion of universe, at the early stage of its existence, exceeded the speed of light.

Another commonly used interpretation of a light cone is that the upper part contains events that the observer can influence, and the lower part contains events from the past that can affect the observer.

Although the general theory of relativity has received experimental confirmation and acceptance, the very fact of considering time as a coordinate qualitatively similar to three spatial coordinates is not satisfactory from a philosophical point of view for many scientists. It is also a generally recognized discrepancy that time is considered to have a *direction* (as opposed to space). This property of time is called its anisotropy.[7]

The fact that time is directed (as opposed to the isotropic nature of space) found its expression in the expression: *arrow of time*. There are three main types of arrows of time:[8]

1. Thermodynamic arrow of time: the direction of time in which entropy (a measure of randomness) increases.
2. Psychological arrow of time: the direction in which we feel the flow of time, the direction in which we remember the past, but not the future.
3. Cosmological arrow of time: the direction of time in which the universe expands.

David Bohm (1917–1992), a student of J. Robert Oppenheimer, who worked closely with Albert Einstein and authored the famous theory of the holographic universe, promoted a unique analysis. One of Bohm's most revolutionary conjectures is that our tangible, everyday reality is actually an illusion, like a holographic image. And beneath this image there is a deeper order of existence – the infinite and primordial level of reality, from which all objects are born. Bohm calls this deep level of reality *implicative* (hidden), while he calls our own level of existence *explicative* (external).

According to Bohm, creation originates from the subtlest layers of the implicative level, where image and reality are indistinguishable. Our automatic division of time into past, present, and future is just a

---

7 Grünbaum A., "Anisotropy of Time." *The Monist* 48 (2):219-247 (1964).

8 One can read about the relationship between different definitions of the arrow of time in the 9th chapter of Stephen Hawking's book, *A Brief History of Time*.

construction of the mind, which "unfolds" from the implicative level; but the implicative level itself does not involve time. This leads into the Vedic concept of the spiritual and material worlds, and their relationship with the energy known as time. This will be discussed in the following sections.

In the middle of the twentieth century, the Soviet scientist N.A. Kozyrev put forward a theory of time that parallels certain Vedic conceptions. While Newton referred to time as entirely passive, Kozyrev acknowledges the active feature of time by highlighting how natural processes are accelerated or slowed down due to its influence.

Interestingly, Kozyrev considered time as perhaps the most important tool used by the Creator of this world, since time not only destroys, but also constantly creates, new natural systems, thereby counteracting the growth of chaos and consequent heat death of the universe. Without rejecting thermonuclear reactions in the interiors of stars, he argued that such fusion processes are clearly not able to account for the colossal reserves of energy needed to sustain the observed radiation levels for most stars, and postulates that time itself may contribute up to 90% of their energy supply.[9] Kozyrev also suggested that the impact of time on the world around us can be characterized by a rotating movement, which corresponds to the Vedic concept of the *kāla-cakra* – the wheel of time. So in this sense he mirrors the Vedic understanding that all processes, from the movement of electrons around an atom to stars around the center of a galaxy, are influenced by the energic feature of time.

Analyzing the results of astronomical observations and laboratory experiments, N. A. Kozyrev concluded that life forms are energized by the spiral design inherent to certain fundamental components. Everyone knows about the spiral shapes of galaxies, the spiral movement of planets, the spiral direction of shell growth, and the spiral structure of DNA and other molecules. This award-winning astrophysicist expressed the idea that the spiral structure of objects and organisms allows them to enhance their life processes by absorbing the flow of time.

To conclude this small historical review, we quote Stephen Hawking from the book, *A Brief History of Time*:

---

9 Note that this approach is consistent with verse 5.52 of the *Brahma-saṁhitā*.

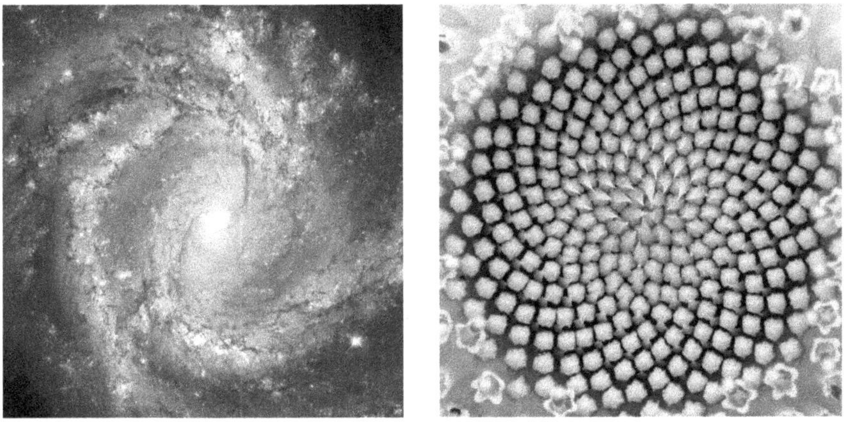

Figure 3: Spiral structures of a galaxy and a sunflower.

Most people would find the picture of our universe as an infinite tower of tortoises rather ridiculous, but why do we think we know better? What do we know about the universe, and how do we know it? Where did the universe come from, and where is it going? Did the universe have a beginning, and if so, what happened before then? What is the nature of time? Will it ever come to an end? Can we go back in time? Recent breakthroughs in physics, made possible in part by fantastic new technologies, suggest answers to some of these longstanding questions. Someday these answers may seem as obvious to us as the earth orbiting the sun—or perhaps as ridiculous as a tower of tortoises. Only time (whatever that may be) will tell...

2. Vedic Concept of Time

As Jīva Goswami states in the *Tattva-sandarbha*, his well-known treatise concerning Gauḍīya Vaiṣṇava epistemology, the most effective means for obtaining knowledge is *śabda:* authoritative testimony coming through a bona fide line of disciplic succession. *Śabda* (also called *avaroha-panthā*, deductive method) generally takes precedence over *pratyakṣa* (experimental, empirical knowledge) and *anumāna* (logical reasoning), without diminishing their importance. Accordingly, the basis for this presentation of the Vedic concept of time will be the evidence from such Vedic scriptures as *Śrīmad-Bhāgavatam* and *Bhagavad-gītā*.

Unlike the approach of modern science, *Śrīmad-Bhāgavatam* and *Bhagavad-gītā* describe time (in Sanskrit - काल, *kāla*) as the active principle behind everything that happens in the material world. Time is an aspect of the localized all-pervading principle of the absolute truth known as Paramātmā,[10] and is also the embodiment of the glance of the Supreme Lord. Time is the instrument for the creation and destruction of the universes, as well as the impetus for the creation, mutual transformation, and combination of primary elements. The cosmology of *Śrīmad-Bhāgavatam* combines the traditional Vedic concepts – *jīva* (living beings, carriers of consciousness), *prakṛti* (material nature), and Īśvara (God, the source of material and spiritual energies) – into a unified structure.

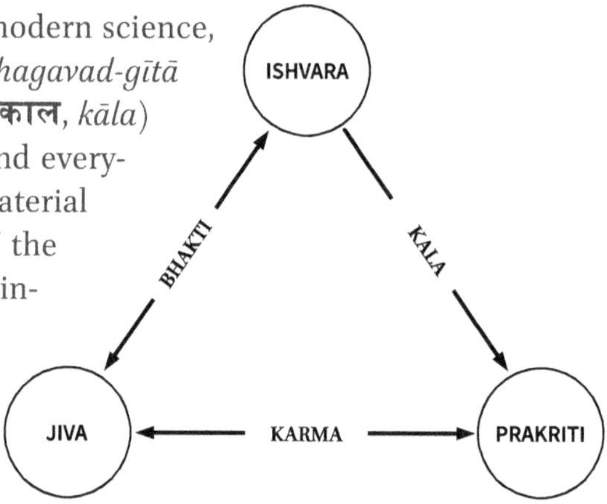

Figure 4: Īśvara, *jīva*, *prakṛti*, *karma*, *kāla*. Arrow from Īśvara to *prakṛti*.

10 For the connection between the energy of time and Paramātmā, see the verses of *Śrīmad-Bhāgavatam* 1.8.28 and 3.26.18.

As Bhaktivedanta Swami Prabhupāda writes in the introduction to *Bhagavad-gītā As It Is*:

> From *Bhagavad-gītā* we must learn what God is, what the living entities are, what *prakṛti* is, what the cosmic manifestation is, how it is controlled by time, and what the activities of the living entities are... Material nature itself is constituted by three qualities: the mode of goodness, the mode of passion, and the mode of ignorance. Above these modes there is eternal time, and by a combination of these modes of nature and under the control and purview of eternal time there are activities, which are called *karma*.

A more detailed description of the time is given in the Third Canto of the *Śrīmad-Bhāgavatam*. In verse 3.10.11 Maitreya says to Vidura:

> Eternal time is the primeval source of the interactions of the three modes of material nature. It is unchangeable and limitless, and it works as the instrument of the Supreme Personality of Godhead for His pastimes in the material creation.

In a purport to this text, Bhaktivedanta Swami Prabhupāda writes:

> The impersonal time factor is the background of the material manifestation as the instrument of the Supreme Lord. It is the ingredient of assistance offered to material nature. No one knows where time began and where it ends, and it is time only which can keep a record of the creation, maintenance and destruction of the material manifestation. This time factor is the material cause of creation and is therefore a self-expansion of the Personality of Godhead. Time is considered the impersonal feature of the Lord.

Thus, time is the instrument through which the Personality of Godhead controls this world. Time is impersonal, since we cannot refer to its "personal" qualities, but we can only say what it is not: unlimited, unchanging, incomprehensible, devoid of variety.

Further, in verse 3.10.12, Maitreya continues:

> This cosmic manifestation is separated from the Supreme Lord as material energy by means of *kāla*, which is the unmanifested, impersonal feature of the Lord.

In a purport to this text, Bhaktivedanta Swami Prabhupāda writes as follows:

> This manifested world is the selfsame Personality of Godhead, but it appears to be something else beyond or besides the Lord. It appears so because of its being separated from the Lord by means of *kāla*. It is something like the tape-recorded voice of a person who is now separated from the voice. As the tape recording is situated on the tape, so the whole cosmic manifestation is situated on the material energy and appears separate by means of *kāla*.

This is in line with what Kṛṣṇa says about His separated energies in verse 7.4 of *Bhagavad-gītā*:

bhūmir āpo 'nalo vāyuḥ
khaṁ mano buddhir eva ca
ahaṅkāra itīyaṁ me
bhinnā prakṛtir aṣṭadhā

> "Earth, water, fire, air, ether, mind, intelligence and false ego – all together these eight constitute My separated material energies."

So time is the factor that separates the energies of Kṛṣṇa from Him, their source; on the contrary, Brahman, the impersonal aspect of Kṛṣṇa, is not His separate part. Also, time "pushes" the *karma* of living beings by initiating the causal mechanisms underlying their activities in the material world. Time determines, or "manifests," the successive states of the universe, making "time travel" impossible. As stated in the commentary to verse 3.11.4 of the *Śrīmad-Bhāgavatam*, each atom is a subtle form of time. Each moment within space-time is characterized by a unique combination of the modes of material nature, which traditionally (in terms of astrology) can be expressed by a particular combination of planets.

3. The Creator's Plan

"When His desire is manifested, it is called time."
Narahari Sarakāra (*Śrī-Kṛṣṇa-Bhajanāmṛta*)

In verse 12.4.37 of *Śrīmad-Bhāgavatam* it is said:

> These stages of existence created by beginningless and endless time, the impersonal representative of the Supreme Lord, are not visible, just as the infinitesimal momentary changes of position of the planets in the sky cannot be directly seen.

So, time, in an imperceptible way for us, implements the Creator's plan in relation to the material world. What is this plan? Stephen Hawking suggests that, at the culmination of its development, modern physics will be able to answer this question: What was God's plan when He created the universe? But the Vedic scriptures provide an answer to the question right from the very start:

1. to enable living beings to imagine they are enjoying independently of God;
2. to facilitate their decision to return to the spiritual world.

Thus, the material world can be viewed as the response to the desires of living entities to realize their freedom of will.

Simultaneously with God's plan for the universe, He also has a plan for every living being. In *Śrīmad-Bhāgavatam* 1.9.14–16, Bhīṣmadeva speaks these words to Mahārāja Yudhiṣṭhira:

> In my opinion, this is all due to inevitable time, under whose control everyone in every planet is carried, just as the clouds are carried by the wind. Oh, how wonderful is the influence of inevitable time! It is irreversible – otherwise, how can there be reverses in the presence of King Yudhiṣṭhira, the son of the demigod controlling religion; Bhīma, the great fighter with

> a club; the great bowman Arjuna with his mighty weapon Gāṇḍīva; and above all, the Lord, the direct well-wisher of the Pāṇḍavas? O King, no one can know the plan of the Lord [Śrī Kṛṣṇa]. Even though great philosophers inquire exhaustively, they are bewildered.

In his commentary on these texts, Bhaktivedanta Swami Prabhupāda writes:

> There is control by time all over the space within the universe, as there is control by time all over the planets. All the big gigantic planets, including the sun, are being controlled by the force of air, as the clouds are carried by the force of air. Similarly, the inevitable *kāla,* or time, controls even the action of the air and other elements. Everything, therefore, is controlled by the supreme *kāla,* a forceful representative of the Lord within the material world. Thus Yudhiṣṭhira should not be sorry for the inconceivable action of time... The Pāṇḍavas suffered so many practical reverses, which can only be explained as due to the influence of *kāla,* inevitable time. *Kāla* is identical with the Lord Himself, and therefore the influence of *kāla* indicates the inexplicable wish of the Lord Himself. There is nothing to be lamented when a matter is beyond the control of any human being... Bhīṣma wanted to impress upon Mahārāja Yudhiṣṭhira that since time immemorial no one, including such demigods as Śiva and Brahmā, could ascertain the real plan of the Lord. So what can we understand about it? It is useless also to inquire about it. Even the exhaustive philosophical inquiries of sages cannot ascertain the plan of the Lord. The best policy is simply to abide by the orders of the Lord without argument. The sufferings of the Pāṇḍavas were never due to their past deeds. The Lord had to execute the plan of establishing the kingdom of virtue, and therefore His own devotees suffered temporarily in order to establish the conquest of virtue.

It is also appropriate here to quote from a letter from Bhaktivedanta Swami Prabhupāda to his disciple Tejiyas on December 19, 1972:

"Krishna has got some plan for you, always think in that way, and very soon He will provide everything to your heart's desire."

So, God has a plan for every living being (soul). Accepting this plan and acting in accordance with it is the proper behavior for the living entities in the material world.

4. The Role of Time in the Creation of the Universe

"God may know how the universe began, but we cannot give any particular reason for thinking it began one way rather than another."
Stephen Hawking (A Brief History of Time)

Thus, *Śrīmad-Bhāgavatam* gives the answer to the question about the cause of the creation of the material world (the question "why?"). But in no less detail, this scripture sheds light on the "technical" side of the question of creation (the question "how?"). This question is described from different perspectives in different places in the *Bhāgavatam*: the third chapter of the First Canto, the fifth chapter of the Second Canto, and the fifth and twenty-sixth chapters of the Third Canto. There is a difference in some details of the description of the process of creation by various sages. This is to be expected since the Absolute Truth is multifaceted and is revealed depending on one's relationship with it. However, all descriptions agree that time is the personification of the glance of the Supreme Personality of Godhead, an impulse that agitates the balance of unmanifested matter (*pradhāna*), as a result of which the three modes of material nature (goodness, passion, and ignorance) appear in *pradhāna*, and then the gradual formation of subtle and gross material elements begins.

Note that such a description does not formally contradict the modern theory of the Big Bang. In his commentary on text 9.5.5 of the *Śrīmad-Bhāgavatam*, Bhaktivedanta Swami Prabhupāda writes:

> Western philosophers sometimes think that the original cause of creation was a chunk that exploded. If one thinks of this chunk as the total material energy, the *mahat-tattva,* one can understand that the chunk was agitated by the glance of the Lord, and thus the Lord's glance is the original cause of material creation.

Further, under the influence of the time factor, there is a sequential transformation of one element into another. Five basic gross elements (*mahā-bhūtas*) – ether, air, fire, water, and earth – are formed from one another under the influence of the time factor. They also inherit the qualities of the elements that precede them.

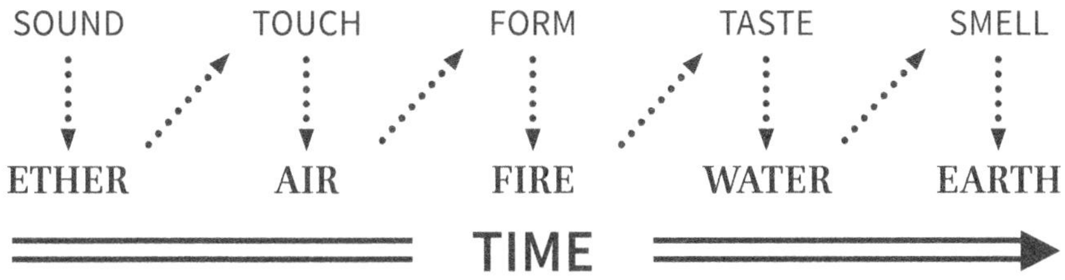

Figure 5: Sound/ether, touch/air, form/fire, taste/water, smell/earth

Note that the appearance of each primary element is preceded by the appearance of its subtle form (sound for ether, touch for air, form for fire, taste for water, smell for earth).

As Bhaktivedanta Swami Prabhupāda writes in his commentary to SB 2.5.25, each of the subtle elements is related to its gross analogue in the same way as "the seer and the seen," which provides a kind of analogy with modern quantum mechanics. Note that there is no unambiguous analogy between the five primary elements listed here and the elements of the modern periodic table of Mendeleev, or the elementary particles currently accepted in Standard Model of particle physics. Rather, these elements are more subtle in nature, and are often categorized as "energies" or "forces": ether provides space; air is a factor of movement; fire provides form; water provides binding interactions, mixing; earth is connected with the mass and the structure of matter.

5. Destructive Action of Time

Although the activity of time is associated with the stages of creation, maintenance, and destruction, the main emphasis is generally placed on its destructive aspect. In verse 9.8 of *Bhagavad-gītā*, Kṛṣṇa says:

> The whole cosmic order is under Me. Under My will it is automatically manifested again and again, and under My will it is annihilated at the end.

After witnessing the world's first nuclear explosion on July 16, 1945, over the New Mexico desert, and being amazed by the spectacle, one of its creators, Robert Oppenheimer, quoted verse 11.32 of the *Bhagavad-gītā*: "Now I am become Death, the destroyer of worlds." (*kālo 'smi loka-kṣaya-kṛt pravṛddhaḥ*).

Chapter 4 of the Twelfth Canto of *Śrīmad-Bhāgavatam* speaks of several types of destruction: *naimittika* (destruction of planetary systems at the end of each *kalpa*), *prākṛtika* (all primary elements under the influence of time are disassembled into their components, lose their qualities, and completely merge with each other), and imperceptible (destruction that occurs constantly under the influence of time).[11] In verse 12.4.36 of the *Bhāgavatam* it is said:

> All material entities undergo transformation and are constantly and swiftly eroded by the mighty currents of time. The various stages of existence that material things exhibit are the perpetual causes of their generation and annihilation.

Further, in the purport to verse 12.4.37 by the disciples of Bhaktivedanta Swami Prabhupāda, it is said: "Time, the potency of the Lord, is very subtle and powerful and is an insurmountable barrier to fools who are trying to exploit the material creation."

---

11 Also, *Śrīmad-Bhāgavatam* mentions the fourth type of destruction, *mukti* – the liberation of a living entity from gross and subtle material identifications.

Time destroys by nature. But this destructive aspect, generally attributed to the mode of ignorance, is in fact associated with the activities of living beings in their attempts to exploit the resources of material nature (*Bhagavad-gītā* 7.5). Thus the energy of time, implementing the plan of God, destroys everything that is perishable, leaving the true values, which have stood the test of time. Everything is "returning on its course" (Ecclesiastes 1.6) and our "great" plans will very soon sink into oblivion, but time will continue to move, reminding living beings of the need to embark on a spiritual path. In a commentary to text 3.26.16 of *Śrīmad-Bhāgavatam*, Bhaktivedanta Swami Prabhupāda writes:

> In other words, time is destructive. Whatever is created is subject to destruction and dissolution, which is the action of time. Time is a representation of the Lord, and it reminds us also that we must surrender unto the Lord. The Lord speaks to every conditioned soul as time.

6. The Measurement of Time

"O inaugurator of the material energy, this wonderful creation works under the control of powerful time, which is divided into seconds, minutes, hours and years."
Śrīmad-Bhāgavatam 10.3.26

Is it possible to measure time? In a commentary to verse 3.10.11 of the *Śrīmad-Bhāgavatam*, Bhaktivedanta Swami Prabhupāda explains that time is both absolute and relative.[12] When we talk about absolute time, we mean the energy that drives the universe. As for relative time, we generally think of it as something that can be measured. Actually, we cannot measure time directly, but we *can* measure its influence "in relation to the speed, change and life of a particular object" (SB 3.10.11, purport). The construction of any clock – be it mechanical, sand, or atomic – is based on this principle. For instance, atomic clocks depend on the frequency of electromagnetic radiation at the atomic level and are indispensable in GPS navigation systems.

One of the leading astrophysicists of our time, Neil deGrasse Tyson, speaks in a similar way about the possibility of measuring time: "Measuring time is possible only through observation of periodic processes."

Which of the periodic processes is the most accessible to us? In the traditional Vedic system, the measurement of time is based on the movement of the sun. In this regard, in verse 3.11.15 of *Śrīmad-Bhāgavatam*, Maitreya says to Vidura:

> O Vidura, the sun enlivens all living entities with his unlimited heat and light. He diminishes the duration of life of all

12 Interestingly, a classification of the types of time, similar to the explanation of Bhaktivedanta Swami Prabhupāda, is given by Isaac Newton in *Philosophiæ Naturalis Principia Mathematica* (1687): "Absolute, true and mathematical time, of itself and from its own nature, flows equably without regard to anything external, and by another name is called duration. Relative, apparent and common time is some sensible and external . . . measure of duration by the means of motion, which is commonly used instead of true time."

> living entities in order to release them from their illusion of material attachment, and he enlarges the path of elevation to the heavenly kingdom. He thus moves in the firmament with great velocity.

In the *Bhāgavatam*, Canto 3, Chapter 11, "Calculation of Time, from the Atom," the issue of measuring time is discussed. A progressive scale of time units is delineated, beginning from the minute intervals involved in atomic processes and gradually progressing to the vast periods involved in the cyclic existence of the universe.

Figure 6: Brihat Samrat Yantra (Jaipur) – the world's largest sundial.

As in the general theory of relativity developed in the twentieth century by Albert Einstein, in the *Bhāgavatam* space and time are considered as closely interrelated categories ("Time and space are two correlative terms," *Śrīmad-Bhāgavatam,* 3.11.4, purport). In the *Purāṇas*, the measurement of time is based on space, which is occupied by particles of matter. One of these particles of matter in Sanskrit is called *paramāṇu* (literally "ultra-small") and the same name *paramāṇu* is assigned to the unit of time during which the sun covers the distance corresponding to one particle of *paramāṇu*. Six *paramāṇus* form one *trasareṇu* particle, which is compared to a particle of dust that can be seen in a sunbeam entering a room. Interestingly, according to *Bṛhat-saṁhitā* and *Matsya Purāṇa,* the size of the *trasareṇu* is 1/8 of the *bālāgra* (the size of the tip of the hair) and is the modern equivalent to about 1 micron (1/1000 of a millimeter).

The time required for the integration (combination) of three *trasareṇu* particles is called *truṭi,* which is 8/13,500 seconds. Further, in the verses of *Śrīmad-Bhāgavatam* there is a recursive chain of relationships that establishes a relationship between other units of time:

100 *truṭis* = 1 *vedha* (8/135 seconds)

3 *vedhas* = 1 *lava* (8/45 seconds)[13]

3 *lavas* = 1 *nimeṣa* (8/15 seconds)[14]

3 *nimeṣas* = 1 *kṣaṇa* (8/5) seconds)[15]

5 *kṣaṇas* = 1 *kāṣṭha* (8 seconds)

15 *kāṣṭhas* = 1 *laghu* (2 minutes)

15 *laghu* = 1 *daṇḍa* (30 minutes)

6 *daṇḍas* = 1 *prahara* (3 hours)

Further, according to verse 3.11.10, eight *praharas* constitute one solar day, which allows the aforementioned modern equivalents for *truṭi* and other units of time to be identified in a proportional way.

We can notice that the Vedic values that characterize the integration time of the particles are significantly larger in comparison with the values given by modern science. In this regard, it should be noted that the concept of *trasareṇu* is different from the modern concept of the atom. The size of a *trasareṇu* particle is significantly larger than the size of an atom, although it is comparable to the size of complex protein molecules. But the very principle of using atomic processes to

---

13 According to the *Śrī Caitanya-caritāmṛta, Madhya* 22.54, one *lava* (0.18 seconds) performed in the company of saintly persons is sufficient to achieve perfection in life.

14 *Nimeṣa* literally means "moment" or time to blink an eye.

15 In Sanskrit poetry, the word *kṣaṇa* is often used as a "moment in time" (see, for example, the fifth verse of "Śrī Gurvaṣṭaka" by Viśvanātha Cakravartī Ṭhākura).

measure time should be the focus here. This is exactly the approach that modern science uses. For example, one second as a unit of time is defined currently as the time during which the cesium-133 atom makes 9,192,631,770 transitions between energy levels.

7. Measurement of Long Periods of Time

So, we have considered the basic units of time corresponding to micro-processes. Likewise, the *Purāṇas* provide a stock of information about units of time for measuring relatively long periods. Interestingly, all *Purāṇas* agree on the number of years in each time unit. The durations of the major time periods are shown in Figure 7.

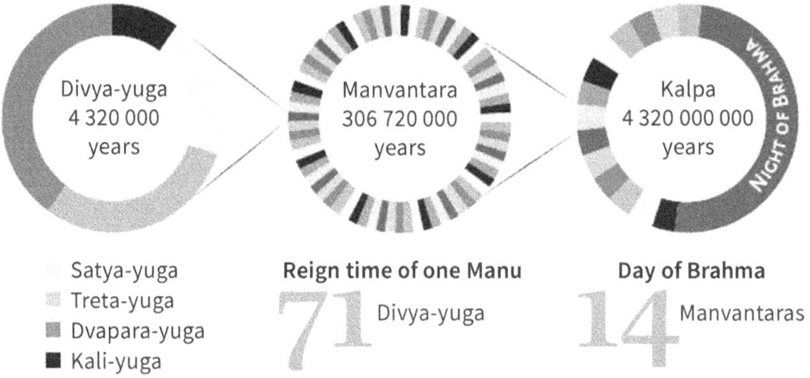

Figure 7. Satya-yuga, Tretā-yuga, Dvāpara-yuga, Kali-yuga. Time for one Manu, Kalpa, and day of Brahmā.

Among the epochal Purāṇic time units, the smallest is Kali-yuga with a duration of 432,000 years, which is preceded by Dvāpara-yuga (864,000 years), Tretā-yuga (1,296,000 years) and, finally, Satya-yuga (1,728,000 years). The cycle of these four *yugas* is called a *cātur-yuga* or *divya-yuga* (4,320,000 years). Seventy-one *cātur-yugas* constitute a *manvantara*. The duration of one *manvantara*, taking into account the *manvantara-sandhi* (transition period lasting 1,728,000 years), is 308,448,000 years.

Fourteen *manvantaras* and fifteen intermediate periods between them constitute one *kalpa* (day of Brahma):

$$14 \times 71 \times 4{,}320{,}000 + 15 \times 1{,}728{,}000 = 4{,}320{,}000{,}000 \text{ years.}$$

And Brahmā's night (the period between the partial destruction of the universe and the new cycle of its intermediate creation) has the same duration. Brahmā lives for 100 × 360 = 36,000 of such "days," which is the lifetime of the universe (311,040,000,000,000 earth years).

8. Our Position in Vedic Chronology

Next, we will consider our current position in the Purāṇic chronology and make a comparison with the data of modern science.

The last mass extinction of species

The *Śrīmad-Bhāgavatam* explains that a devastating flood, inundating the earth and causing almost complete destruction of its inhabitants, occurs at the end of each *manvantara*.

According to the *Vāyu Purāṇa* (1.23.111–207), the twenty-eighth *cātur-yuga* of the seventh *manvantara* (Vaivasvata *manvantara*) is now in progress. In accordance with astronomical calculations (one can read more about them in the book by Richard L. Thompson, *Vedic Astronomy and Cosmography*), Kali-yuga started in 3102 BC. The great Indian mathematician Āryabhaṭa reported that he was 23 years old when 3600 years of Kali-yuga had passed. Since Āryabhaṭa was born in 476 AD, this confirms a similar date for the beginning of Kali-yuga. To simplify the calculations, we will assume that 5000 years of Kali-yuga have passed at the moment. Then, accordingly, 28 × 4,320,000 - 432,000 + 5,000 = 120,533,000 years have passed since the beginning of the current *manvantara*.

To compare, the dates of the mass extinctions at the end of the Triassic and Cretaceous periods occurred, according to modern science, about 200,000,000 and 65,000,000 years ago respectively.

The formation of the solar system

From the Vedic perspective, the formation of planets, stars, and other celestial objects refers to the beginning of a *kalpa* (day of Brahmā). We have already mentioned that the current *manvantara* is the seventh out of fourteen. Thus, our position relative to the beginning of *kalpa* is 120,533,000 + 6 × 308,448,000 = 1,971,221,000 years. If we compare this time with the time of formation of the solar system from the "protosolar" molecular cloud (4,567,300,000 years ago) as per modern calculations, we will see that these values have the same order of magnitude.

The lifetime of the universe

According to the eleventh chapter of the Third Canto of *Śrīmad-Bhāgavatam*, we are currently in the first day of the second half of Brahmā's life. From this it can be approximately calculated that the time of our universe at the moment is 155,521,971,221,000 years. The modern estimate of the age of the universe (from the moment of the Big Bang) is 13,801,000,000 years. Note that in his book, *The Cycles of Time*, Roger Penrose, winner of the 2020 Nobel Prize in Physics, lays out the concept of a cyclic universe, which is reminiscent of the cyclic nature of time described in the *Vedas* and does not limit the beginning of the universe to the moment of the Big Bang.

One of the prominent astrophysicists of the twentieth century, Carl Sagan, said that modern theories about the cyclical degeneration of the cosmos over long periods of time find surprising similarities with knowledge from ancient Indian manuscripts. Could this be just a coincidence? We will leave the answer to this question to the reader.

9. Time at Different Planetary Levels

Śrīmad-Bhāgavatam and other *Purāṇas* speak of various time scales at different planetary levels (*lokas*). These data are shown in the table.

Cosmic level	Lifespan (in earthly years)	Universe's (in years) at the given level
Brahmaloka	311.04×10^{12}	100
Maharloka	4.32×10^{9}	7.2×10^{9}
Svargaloka (Indra, Manu)	306.7×10^{6}	101.5×10^{6}
Candraloka	36,000	864×10^{9}
Earthly level	100	311.04×10^{12}

In verse 3.11.12 of *Śrīmad-Bhāgavatam* it is said that every living entity has a lifespan of 100 years. However, the duration of these 100 years is different for each level of the planetary systems. This is confirmed by a number of stories from the *Śrīmad-Bhāgavatam* and other *Purāṇas*. Thus, in the story of King Kakudmī from the Ninth Canto of *Śrīmad-Bhāgavatam*, it is said that during a short concert in Brahmaloka, performed by the Gandharvas, 27 *cātur-yugas* passed on Earth. Considering that one "second" on Brahmaloka is 8,640,000,000/24/3600 = 100,000 earth years, the period of 27 earthly *cātur-yugas* is 27 × 4,320,000/100,000 = 1166 "seconds" ≈ 20 "minutes" for Brahmā and is an adequate period of time for a small concert.

Similar stories are given in the thirteenth chapter of the Tenth Canto of *Śrīmad-Bhāgavatam*, where in the fortieth verse it is mentioned that one earthly year had passed during one moment for Brahmā, as well as the story of Gopa-kumāra in *Bṛhad-bhāgavatāmṛta* by Sanātana Gosvāmī: during the short time spent by the hero on Brahmaloka, many epochs had passed on Earth. In this regard, we note that a full-fledged visit by a person on a particular planet is possible only when they begin to live according to the time cycle of that planet.

Note that Einstein's general theory of relativity also predicts time dilation for objects moving at a speed close to the speed of light.[16] Interestingly, from a geocentric perspective, the planets make one revolution around the earth in 24 hours. A planet located near the shells of Brahmāṇḍa (Purāṇic universe), that is, at a distance of 250,000,000 *yojanas* (3.2 billion kilometers or 2 billion miles) from the earth, would have a rotation speed $v = 2\pi \times 3.2 \times 10^9/24/3600 = 232{,}710$ km/s (144,600 mi/s), which is comparable to the speed of light ($c = 299{,}792$ km/s or 186,282 mi/s). Of course, these arguments are not rigorous and are presented here as an example for establishing possible correspondences between the Vedic and modern Western concepts of space-time.

Along with the description of various time scales, it would be appropriate to speak about the effect of relativity of the perception of time. Each of us knows from our own experience that the passage of time is perceived differently at different ages and under different circumstances. As is mentioned by the author of the *Theory of Time,* professor N. A. Kozyrev, with the aging of a person "the density of time decreases."

Our perception of time is also affected by the particular influence of the *guṇas* (modes of nature) on us at the moment. As all colors are formed by the combination of red, green, and blue, so every moment of space-time is characterized by a unique combination of the modes of material nature. As Bhakti Vijnana Goswami writes in his essay "The Secret of Eternal Life," the research of modern scientists has shown that subjectively, for ordinary people, the present lasts no more than a few seconds. The stronger the *guṇa* of passion, the "shorter" the present becomes, squeezed from both sides by the past and the future (the burden of the past and hopes for the future). People in the mode of goodness, however, have the ability to gradually "expand" the present, and this ability is proportional to the strength of the influence of the mode of goodness. The mode of goodness (*sattva-guṇa*) slows down time. Yogis deliberately slow down their breathing in order to stop the endless rushing of the mind. In this stream of thoughts, flowing endlessly from the past to the future, we do not notice the present. But when the mind slows down, the passage of time slows down along with it, and this is one of the reasons why yogis live longer. All physiological

16 The famous "twins paradox" suggests that in order to age 1.5 times slower, you need to move at a speed of about 90% of the speed of light.

processes in the body of a yogi are also changed qualitatively. When a yogi reaches a state of *samādhi*, the boundaries of the present expand to hours, days, weeks, months, and even years. His consciousness passes into a different realm, and when *sattva-guṇa* becomes pure, completely free from the influence of *rajo-guṇa* (passion) and *tamo-guṇa* (ignorance), the present becomes eternity.

10. Time in the Spiritual World

"What is the ultimate scope of science? Is it just the material attributes of our universe that are amenable to its methods?"
Roger Penrose (*Shadows of the Mind*)

As can be seen from the table of time scales on various planetary systems (see chapter 9), the higher we move away from the level of our planet (approaching the shells of the material universe), the more "fleeting" time is. The lifetime of the universe is only one hundred years for Brahmā and fits into just one cycle of Mahā-Viṣṇu's breathing. From this we can conclude that outside the material world there is no flow of time in the sense we are accustomed to; that is, there is no place for a mechanistic approach to the law of cause and effect.

In a commentary to text 2.9.10 of *Śrīmad-Bhāgavatam*, Bhaktivedanta Swami Prabhupāda writes:

> In the material world everything is created, and everything is annihilated, and the duration of life between the creation and annihilation is temporary. In the transcendental realm there is no creation and no destruction, and thus the duration of life is eternal unlimitedly. In other words, everything in the transcendental world is everlasting, full of knowledge and bliss without deterioration. Since there is no deterioration, there is no past, present and future in the estimation of time. It is clearly stated in this verse that the influence of time is conspicuous by its absence. The whole material existence is manifested by actions and reactions of elements which make the influence of time prominent in the matter of past, present and future. There are no such actions and reactions of cause and effects there, so the cycle of birth, growth, existence, transformations, deterioration and annihilation — the six material changes — are not existent there.

Note that the concept of an ideal world, the imperfect reflection of which is the material world around us, is not at all alien to modern scientists. To varying degrees, the concept of an ideal (Platonic) world was taken as a basis by Karl Popper, Roger Penrose, and Arnold Neumeier.

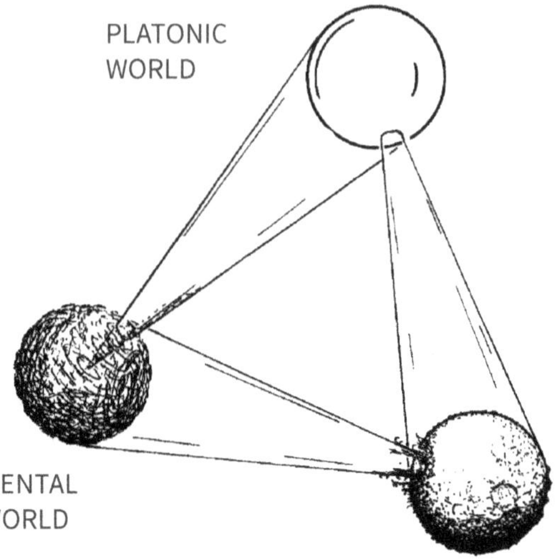

Figure 8: The concept of the physical, mental, and Platonic worlds proposed by Roger Penrose.

As Dr. Thoudam D. Singh (Bhaktisvarūpa Dāmodara Mahārāja) notes,[17] the description of the action of time in the spiritual world has a conceptual similarity to the concept of imaginary time, introduced by the famous astrophysicist of modern science, Stephen Hawking. In his book, *Black Holes and Baby Universes*, Hawking writes:

> Quantum theory introduces a new idea, that of imaginary time. Imaginary time may sound like science fiction… But nevertheless, it is a genuine scientific concept. One can picture it in the following way. One can think of ordinary, real time as a horizontal line. On the left, one has the past, and on the right, the future. But there's another kind of time in the vertical direction. This is called imaginary time, because it is not the kind of time we normally experience. But in a sense, it is just as real, as what we call "real time."

17 Bhaktisvarūpa Dāmodara Mahārāja, *Life and Origin of the Universe* (Kolkata, India: Bhaktivedanta Institute, 2004).

So, the axis of imaginary time is perpendicular to the axis of "real" time, just as the axis of imaginary numbers in mathematics is perpendicular to the axis of real numbers, as is shown in the figure below.

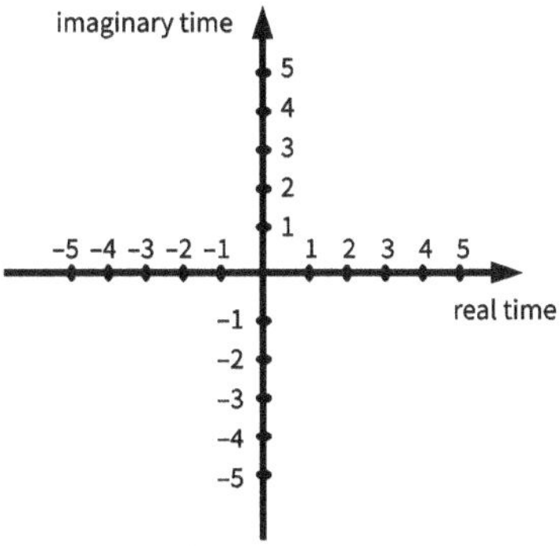

Figure 9: Real time, imaginary time

According to Hawking:

> This might suggest that the so-called imaginary time is really the real time, and that what we call real time is just a figment of our imaginations. In real time, the universe has a beginning and an end at singularities that form a boundary to space-time and at which the laws of science break down. But in imaginary time, there are no singularities or boundaries. So maybe what we call imaginary time is really more basic, and what we call real is just an idea that we invent to help us describe what we think the universe is like.

11. The Cyclical Nature of Time and a Look at World History

In Chapter 1, devoted to a historical overview, we have already talked about the concept of the "arrow of time." Western science and culture emphasize the linearity of time, that is, its one-pointedness from the past to the future. On the contrary, the Vedic view of the nature of time speaks, first of all, of its *cyclical* nature (without denying, however, the concepts of "past" and "future" inherent to the material world). Note that the difference in these concepts radically affects a person's worldview and spiritual values.

The concept of linear time implies the uniqueness of what is happening in the world as a whole and for each individual in particular. According to the famous Romanian philosopher and writer Mircea Eliade, author of the book *The Myth of the Eternal Return*, the concept of linear time arose relatively recently in the era of the Old Testament (about 5000 years ago) and is characteristic of the Abrahamic religions (according to the *Vedas*, Kali-yuga, the era of degradation of society, began about 5000 years ago). Later, the concept of linear time was developed in the writings of the 17th/18th century German philosophers, Leibniz (1646–1716) and Goethe (1749–1832). Specifically, Leibniz spoke of "the endless progress of man and the material world as a whole."

Although in the twentieth century the development of the Big Bang theory and the "discovery" of the expansion of the universe served to strengthen the position of those embracing the linearity of time, later one of the authors of the Big Bang theory, Roger Penrose, in the previously mentioned book, *The Cycles of Time*, introduced the concept of cycles as an operating principle of the universe: at the end of each cycle the universe is destroyed and then created again.

The concept of time cycles was originally inherent to many world civilizations (Ancient Egypt, Mesopotamia, Greece, India). Everywhere in nature, we see periodic processes: the movement of electrons around an atom and planets around the sun, the change of seasons, and the process of reincarnation. If we accept the idea of a cyclic universe, then what is happening to us and to the world as a whole may not be something completely unique. The cycles of creation and destruction may have occurred an infinite number of times in the past and may also happen again in the future. This naturally prompts persons to seriously consider questions about the significance of life and the purpose of their existence.

The differences in these two (linear and cyclic) approaches to understanding the nature of time leads to differences in understanding history as a science. Historians note a special approach to the way in which history has been preserved and presented in the Indian sub-continent. The historical understanding of India's past generally is based on the *Itihāsas* (which include, first of all, the *Rāmāyaṇa* and *Mahābhārata*), as well as the *Purāṇas*, which were transmitted orally through a chain of disciplic succession and then recorded about 5000 years ago. Bhaktivedanta Swami Prabhupāda explained in a lecture (*Śrīmad-Bhāgavatam* 6.1.16–20, August 2, 1971) in New York City:

> So these *Purāṇas* are explanation of the Vedic system in understandable historical references. Therefore they are called *itihāsam. Itihāsam* means old history. So *Itihāsa* does not mean this has to be learned chronological with date. Now if you keep a history of millions and trillions of years, it is impossible to keep. Therefore most important incidences in the history, they are picked up, and they are assorted in the *Purāṇas*.

Thus, Vedic culture does not necessarily give primary importance to the chronological sequence of specific events. In the Vedic tradition, the author of a literary or scientific treatise generally acknowledges the name of his teacher, the line of succession to which he belongs, influential patrons, the person to whom the treatise was created, and his worshipable Deity. History, from the point of view of Vedic culture, is not a history of events *per se*, but rather a description of a sequence of personalities that suggests a continuity of dynasty, culture, and/or tradition. For example, when the *Purāṇas*, or *Itihāsa*s in general, refer to a "*manvantara*," the period of the reign of one Manu (308,448,000 years), emphasis is placed on of the chain of personalities, members of great dynasties of righteous rulers of the Earth, who reigned during that particular *manvantara*.

Not only in the Vedic scriptures, but also in the Bible and the Koran, the same principle is used: a description of ancestors interspersed with a description of the qualities of those individuals. This suggests that the various world traditions and cultures, regardless of religion or nation, may be traced back to a common, ancient center of civilization. Measuring the passage of time simply based on events, wars, and so on is a sign of impersonalism, which is a characteristic of the age of Kali. On the contrary, Vedic history, as presented in the *Itihāsas*, measures time according to the activities and attributes of great personalities. Personalities make history, not vice versa.

It is interesting that in the famous epilogue to the novel *War and Peace*, Leo Tolstoy rejects the influence of personalities on the course of world history, bringing to the fore "the mysterious forces driving mankind." According to him, "historical figures rush from one coast to another." Citing various historical examples of "great people," he shows how they are puppets in the hands of the forces behind the mechanism of the material world.

But Vedic history describes another class of personalities, many of whom were great saints or divine incarnations, who performed great deeds for the benefit of humanity. By their actions, they changed the course of events, realizing God's plan for the universe. Thus these historical accounts are humanized and imbued with a sense of purpose. This spiritual approach, aimed at developing certain qualities in the reader, educates and guides him by showing how spirit and personality are primary, in opposition to the idea that faceless, inert matter is the fundamental feature of our world.

By understanding the connection between the present generations and those of the distant past, the stories of the *Śrīmad-Bhāgavatam* appear before us as real events associated with our experience of sensory perception, connecting a transcendental reality to a material existence. The genealogy – as chains of heredity, disciplic succession, and dynasties given in the Vedic scriptures – traces back to Lord Brahmā and further to *his* source, Lord Kṛṣṇa, thus establishing Kṛṣṇa as the source of everything in the world.

Vedic history not only tells about the events of the past, but also contains predictions for the future. Such events include, for example, the advents of Lord Buddha (*Śrīmad-Bhāgavatam*, 1.3.24, 2.7.37) and Lord Caitanya (11.5.32-33), which were predicted several thousand years ahead of their appearance. Another characteristic example is found in the first chapter of the Twelfth Canto of the *Śrīmad-Bhāgavatam* (verses 11-12). The verses speak of a certain *brāhmaṇa* who, having betrayed the trust of King Nanda and his eight sons, will destroy their entire family and enthrone King Candragupta. In a commentary to text 11, referring to Śrīdhara Svāmī and Viśvanātha Cakravartī Ṭhākura, it is said that the mentioned *brāhmaṇa* is the famous philosopher and politician Cāṇakya Paṇḍita (350–275 BC), also known as Kauṭilya or Vātsyāyana. Modern historians acknowledge the existence in the past of the Maurya dynasty, as well as the king Candragupta. Thus, the *Śrīmad-Bhāgavatam*, written about 5000 years ago, which begins with the events that preceded the creation of the world, also describes facts about recent history preserved in the memory of mankind.

12. Vedic Calendar

The Vedic culture has developed a sophisticated methodology for timekeeping, calendar making, and for timing the performance of various rituals. In this connection, the typical example given by the the texts of *Śrīmad-Bhāgavatam* is is 7.14.20–23:

> One should perform the *śrāddha* ceremony on the Makara-saṅkrānti [the day when the sun begins to move north] or on the Karkaṭa-saṅkrānti [the day when the sun begins to move south]. One should also perform this ceremony on the Meṣa-saṅkrānti day and the Tulā-saṅkrānti day, in the *yoga* named Vyatīpāta, on that day in which three lunar *tithis* are conjoined, during an eclipse of either the moon or the sun, on the twelfth lunar day, and in the Śravaṇa-nakṣatra. One should perform this ceremony on the Akṣaya-tṛtīyā day, on the ninth lunar day of the bright fortnight of the month of Kārttika, on the four *aṣṭakās* in the winter season and cool season, on the seventh lunar day of the bright fortnight of the month of Māgha, during the conjunction of Maghā-nakṣatra and the full-moon day, and on the days when the moon is completely full, or not quite completely full, when these days are conjoined with the *nakṣatras* from which the names of certain months are derived. One should also perform the *śrāddha* ceremony on the twelfth lunar day when it is in conjunction with any of the *nakṣatras* named Anurādhā, Śravaṇa, Uttara-phalgunī, Uttarāṣāḍhā or Uttara-bhādrapadā. Again, one should perform this ceremony when the eleventh lunar day is in conjunction with either Uttara-phalgunī, Uttarāṣāḍhā, or Uttara-bhādrapadā. Finally, one should perform this ceremony on days conjoined with one's own birth star [*janma-nakṣatra*] or with Śravaṇa-nakṣatra.

Traditionally, calendars in Vedic culture have been based on the movement of the sun and moon relative to the sphere of the stars. The lunar cycle was used to determine the months and days, and the solar cycle was used to determine the full year. The solar year was tracked by observing the entry and exit of the sun into a constellation formed by stars in the sky, which were divided into 12 intervals of 30 degrees each (the transition of the sun from one constellation to another is called *saṅkrānti*). This system is similar to the Hebrew and Babylonian ancient calendars, creating the same problem for accounting for the discrepancy between nearly 354 lunar days in twelve months, compared to over 365 solar days in a year. To eliminate this discrepancy, there were a number of methods for calculating intercalary months (*adhika-māsa* or Puruṣottama-māsa), that is, adding one more month on average every 32.5 months.

Among the important parameters of the Vedic calendar, the main ones were *tithi*, *vāsara*, and *nakṣatra*. Let's consider these parameters in more detail.

1. *TITHI*

A *tithi*, or lunar day, is the time it takes to change the angular distance between the sun and moon by twelve degrees. They begin at different times of the solar day and vary in length from approximately 19 to 26 hours. If more than one *tithi* begins during a solar day, usually the one which occurs during sunrise or, according to the Vaiṣṇava calendar, during the *brāhma-muhūrta* (beginning 96 minutes before sunrise), is considered dominant. The *tithi* number is based on the formula:

tithi = (moon longitude - sun longitude) / 12

and changes from 1 to 30 (if the formula gives the number outside this range, then you need to add 30 or subtract 30 from it). A *tithi* from 1 to 15 corresponds to the light half (*śukla-pakṣa*) of the month, and from 16 to 30 to the dark half (*kṛṣṇa-pakṣa*). The first day of each half is called *pratipadā*. The fifteenth day of the light half is called *pūrṇimā* (full moon), and the fifteenth day of the dark half is called *amāvasyā* (new moon). The rest of the *tithis* of each half of the month are designated in accordance with their ordinal numbers: *dvitīya, trītīya, caturthī, pañcamī, ṣaṣṭhī, saptamī, aṣṭamī, navamī, daśamī, ekadaśī, dvādaśī, trayodaśī, caturdaśī.*

2. *VĀSARA*

Vāsara refers to the day of the week. Traditionally, the beginning of the day was considered to be the moment of sunrise and every hour of the day (*horā*) is ruled by one of the planets. The name of the planet leading every day is determined by the leading planet of the first hour of the given day. This data is given in verse 12.78 of the *Sūrya-siddhānta* and is presented in the table below. Note that the sequence of planets governing each hour is arranged in decreasing order of distance to them from Earth in accordance with the concepts of modern astronomy (Saturn, Jupiter, Mars, Sun, Venus, Mercury, Moon).

time, hours	1	2	3	4	5	6	7
1	Sun	Moon	Mars	Mercury	Jupiter	Venus	Saturn
2	Venus	Saturn	Sun	Moon	Mars	Mercury	Jupiter
3	Mercury	Jupiter	Venus	Saturn	Sun	Moon	Mars
4	Moon	Mars	Mercury	Jupiter	Venus	Saturn	Sun
5	Saturn	Sun	Moon	Mars	Mercury	Jupiter	Venus
6	Jupiter	Venus	Saturn	Sun	Moon	Mars	Mercury
7	Mars	Mercury	Jupiter	Venus	Saturn	Sun	Moon
8	Sun	Moon	Mars	Mercury	Jupiter	Venus	Saturn
9	Venus	Saturn	Sun	Moon	Mars	Mercury	Jupiter
10	Mercury	Jupiter	Venus	Saturn	Sun	Moon	Mars
11	Moon	Mars	Mercury	Jupiter	Venus	Saturn	Sun
…							…
24	Venus	Jupiter	Venus	Saturn	Sun	Moon	Mars

3. *NAKṢATRA*

The ecliptic (the circle of the celestial sphere along which the sun moves) is divided into 27 *nakṣatras* (the modern name is "lunar mansions"). *Nakṣatras* serve, as a rule, to indicate the cycle of motion of the moon relative to the stars. This cycle is approximately 27 days, 7 hours, and 43 minutes.

The division of the ecliptic into *nakṣatras* is made in an easterly direction relative to the star Citra (Spica), the initial star of the Aries constellation. Each *nakṣatra* covers 13°20″ of the celestial sphere; therefore the calculation of the *nakṣatra* is based on the following formula:

nakṣatra = moon longitude from Aries / 13° 20″

The *nakṣatras,* according to their numbers obtained from this formula, have the following names: (1) Aśvinī, (2) Bharanī, (3) Kṛttikā, (4) Rohiṇī, (5) Mṛgaśīrṣa, (6) Ārdrā, (7) Punarasu, (8) Puṣya, (9) Āśleṣā, (10) Maghā, (11) Pūrva-phalgunī, (12) Uttarā-phalgunī, (13) Hasta, (14) Citra, (15) Svāti, (16) Viśākhā, (17) Anurādhā, (18) Jyeśṭhā, (19) Mūla, (20) Pūrvāṣādhā, (21) Uttarāṣādhā, (22) Śravaṇā, (23) Dhaniṣṭhā, (24) Satabhiṣā, (25) Pūrvabhādra, (26) Uttarabhādra, (27) Revatī. The twenty-eighth *nakṣatra*, Abhijit, is placed between the last quarter of the Uttarāṣādhā *nakṣatra* and 1/15 of the Śravaṇā *nakṣatra*.

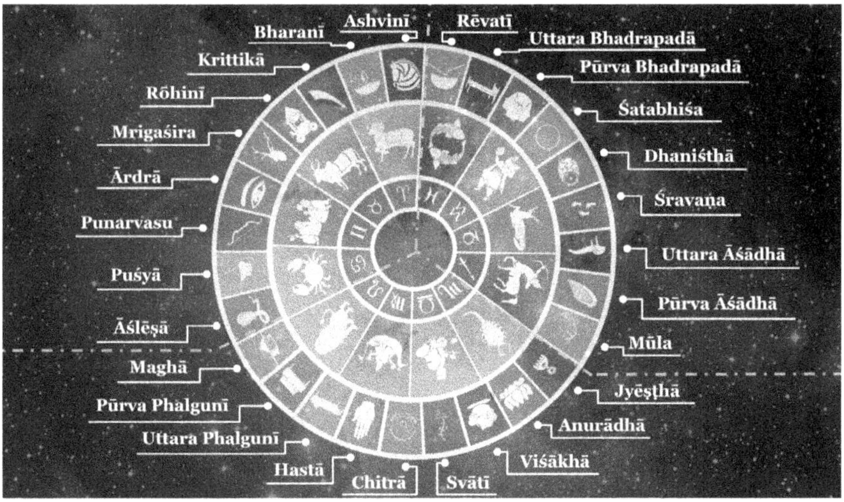

Figure 10: Comparing the *nakṣatras* (outer circle) with the 12 tradition signs of the zodiac (inner circle).

Note that the name of each month is determined by the *nakṣatra* in which the full moon appears that month. For example, if the full moon coincides with the Citra *nakṣatra,* then the corresponding month is called Caitra according to the rules of Sanskrit grammar. The names of the twelve months are as follows: Caitra, Vaiśakha, Jyaiṣṭha, Āṣāḍha, Śrāvaṇa, Bhādra, Āśvin, Kārttika, Margaśirṣa, Pauṣa, Phālguna.

Note that the Vaiṣṇava tradition also uses the Śaka calendar (originating on March 15, 78 A.D.) and Gaurābda calendar (originating from the *pūrṇimā* of the Phālguna month in 1486, the day of the appearance of Śrī Caitanya Mahāprabhu).

13. Time Travel

Although the idea of time travel has been discussed since ancient times, this concept has been recently popularized by the works of science fiction writers, among which the most famous is *The Time Machine* by H. G. Wells (1866–1946), where the hero is transported to the year 802,701 CE.

Einstein's general theory of relativity allows for the existence of paths in space-time that could theoretically be used by time travelers. The well-known Austrian logician, mathematician, and philosopher, Kurt Gödel, famous for his acclaimed incompleteness theorem, proposed a scheme for the theoretical possibility of time travel. According to him, space-time rotation leads to the tilt of light cones[18] for observers A, B, and C, as is shown in Figure 11. If the tilt of the cone is big enough, the future of observer A lies in the past for B, the future of B lies in the past for C, and so on. As a result, the future of observer A "merges" with his past.

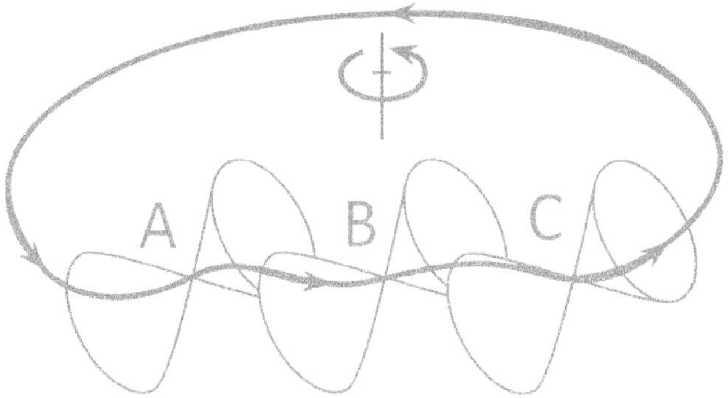

Figure 11. Kurt Gödel's concept of time travel for observers A, B, and C.

This approach by Gödel, though purely theoretical, irritated his friend and colleague, Albert Einstein, but no formal contradictions were found in it.

18 For more information on the light cone, see Chapter 1, " Brief Historical Review."

A more popular theoretical construct (not least because of its name) is the concept of "wormholes" – hypothetical connections between widely separated space-time regions (in their original work of 1935, Einstein and Rosen called these paths "bridges"). Inside the wormhole, one can move from one point in space-time to another due to the focused gravitational fields. But such fields must be very powerful, and they must be controlled with maximum precision.

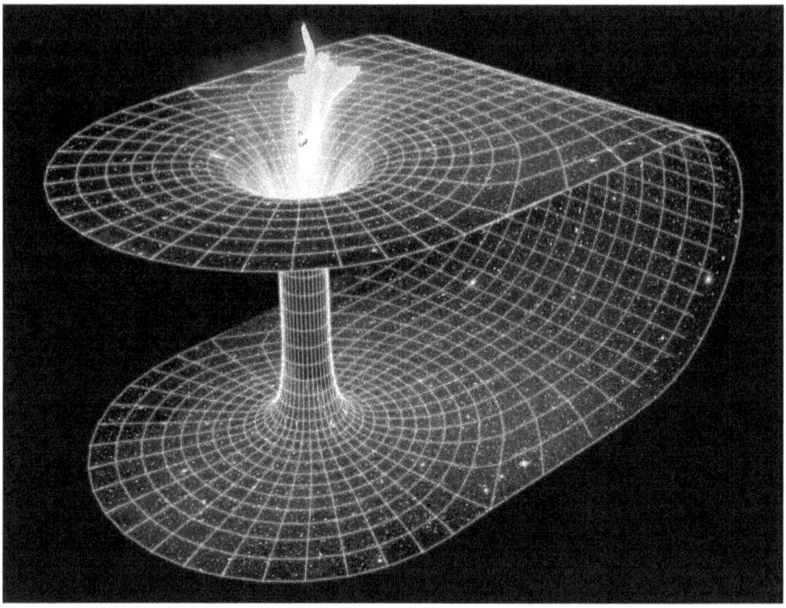

Figure 12. The concept of a wormhole.

Another stumbling block for this theory is the unidirectionality of the wormhole, which means it closes after a single passage through it. The ability to keep a wormhole open requires a negative curvature of space-time (like that of a saddle), while ordinary matter has a positive curvature (like a sphere).

In 2009, Stephen Hawking conducted an interesting "experiment" by organizing a "Time Traveler's Party" at the University of Cambridge and sending out invitations after it was over. But no one from the future came to the party. Hawking explained the absence of visitors by assuming that the past is fixed and does not have the necessary curvature of space-time.

Due to lack of evidence, many modern scientists consider time travel to be practically impossible, for several reasons: 1) They suggest that for the foreseeable future, man-made objects will not be able to approach the speed of light. 2) Also, the huge gravitational field associated with such speeds would be lethal. 3) The requisite deformation of space-time would lead to colossal natural disasters. 4) And another argument, not so scientific but often cited by the scientists themselves, is the "grandfather paradox," where the time traveler returns to the past, kills his ancestor in his youth, and as a result disappears – because he was never born!

There is no mention of time travel into the past in the Vedic literature. Even when Kṛṣṇa Himself descends to the material world to perform His pastimes, He operates within the present conditions, and does not change the past. In the story described in verses 10.45.37–50 of the *Śrīmad-Bhāgavatam,* Sāndīpani Muni asked Kṛṣṇa to return his deceased son to him. Instead of going back to the past, Kṛṣṇa traveled to the abode of the Lord of Death and transported His guru's son back to his home. Thus, Kṛṣṇa did not violate the current understanding that the likelihood of being able to travel into the past is very remote.

However, these complexities associated with time travel do not necessarily exclude the possibility of *knowing* the past and the future. Persons who have this ability are called in Sanskrit *tri-kāla-jñā* (literally "one who knows the three times"). In a commentary to text 4.8.25 of the *Śrīmad-Bhāgavatam,* Bhaktivedanta Swami Prabhupāda writes about Nārada Muni:

> Nārada is *tri-kāla-jña;* he is so powerful that he can understand the past, future and present of everyone's heart, just like the Supersoul, the Supreme Personality of Godhead.

Also in the commentary to text 3.15.3 it is said that "demigods and great sages who have attained such perfection are called *tri-kāla-jña.*"

The *tri-kāla-jña* concept is similar to the phenomenon known today as "precognitive remote viewing." The essence of this concept is as follows: During the experiment, one test person remains in a room with the experimenters and tries to describe what another test person sees at a great distance from him. Thousands of such experiments have been successfully carried out at Princeton University. A surprising feature of the results was that observers could receive information equally well from the past, present, or future.[19] Although the modern, mainstream scientific community is generally not open to evidence supporting instances of individuals receiving accurate glimpses of the past or the future, remote viewing research in the United States, which began with the government-funded Stargate Project in 1978, continues to this day.

19 Richard L. Thompson, "Time Travel and Consciousness," *Back to Godhead* 41, no. 5 (September/October 2007), 46–50.

14. How To Overcome the Influence of Time

"Simply by expansion of His eyebrows, invincible time personified can immediately vanquish the entire universe. However, formidable time does not approach the devotee who has taken complete shelter at Your lotus feet."
Śrīmad-Bhāgavatam 4.24.56

How can a person get out of the influence of time? Verse 2.3.17 of *Śrīmad-Bhāgavatam* states:

> Both by rising and by setting, the sun decreases the duration of life of everyone, except one who utilizes the time by discussing topics of the all-good Personality of Godhead.

Bhaktivedanta Swami Prabhupāda writes in the purport to this text:

> The Lord is the complete spirit whole, and His name, form, quality, pastimes, entourage and personality are all identical with Him. Once a person comes into contact with any one of the above-mentioned energies of the Lord through the proper channel of devotional service, the door to perfection is immediately opened... And thus the sun fails to rob the pure devotee of his duration of life, inasmuch as he is constantly busy in the devotional service of the Lord, purifying his existence. Death is a symptom of the material infection of the eternal living being; only due to material infection is the eternal living entity subjected to the law of birth, death, old age and disease.

In this regard, an analogy with the well-known riddle of Einstein is appropriate. The essence of it is that one can go beyond the circle without crossing its border only by using the third dimension, i.e. height. Likewise, it is possible to go outside the sphere without crossing its border using the fourth dimension. Similarly, one can go beyond the influence of time by using an additional dimension.

Bhaktivedanta Swami Prabhupāda spoke about such a spiritual dimension at a lecture in Mayapur on March 10, 1976, calling this dimension *turya* (literally "fourth"). On the one hand, the fourth means "lying outside the three-dimensional world," and on the other, "different from the three states of being – wakefulness, sleep, and deep sleep."

15. Is Everything Predetermined?

The idea of time as a driving force manifesting the material creation may seemingly lead to complete predetermination (philosophy of determinism): if the initial state of the universe is given, then all its subsequent development is predetermined by the laws established by physics or God. Similarly, in connection with the causal law of *karma*, the question arises: if a living being is dependent on the results of his actions in the past, where is the place for free will? Once committing a sin, a person falls into an endless chain of suffering and becomes an endless victim of the consequences of their actions.

Regarding the predetermination of the fate of the universe as a whole, in his article "The Clockwork Universe in Chaos," Richard L. Thompson mentions three concepts of God's interaction with His Creation:

- Constant direct intervention;
- Pre-programmed history of the universe;
- The functioning of the universe in accordance with a predetermined set of physical laws (the concept of deism).

Isaac Newton, who was a proponent of the first of these concepts, argued that the planets could not move in their orbits due to the inherent instability of their trajectories, and therefore the intervention of God is a necessary factor for the stability of their motion. Note that it is the energy of time that is the instrument with which God controls the material world. The adherents of the opposite, deterministic approach were Leibniz (a constant opponent of Newton), as well as the French scientists Joseph-Louis Lagrange and Pierre-Simon Laplace. "Laplace's demon," a fictional, intelligent creature capable of perceiving at any given time the position and speed of each particle in the universe and exactly predicting its evolution both in the future and in the past, still plays a role in modern scientific discussions concerning determinism, although recently several mathematicians have discovered logical contradictions in the concept.

Interestingly, a number of doubts about predicting events based on some such universal theory were expressed by one of the original well-known adherents of this approach, Stephen Hawking. In his *A Brief History of Time*, Hawking notes:

> 1. Such a theory is supposed to be compact and elegant from the point of view of mathematics. There must be something special and simple in the theory of everything. And yet, how can a certain number of equations take into account all the complexity and the smallest details of what we see around? Could a unified theory determine that Sinéad O'Connor will be the first on this week's hit parade or that Madonna will appear on the cover of *Cosmopolitan*?
>
> 2. Even the equations describing the motion of an electron around a hydrogen atom are too complicated for their direct solution. What can we say then about the prediction in all the details of the fate of the universe?
>
> 3. All our statements and guesses about the great unified theory are also predetermined by this very theory. But why should it predetermine that we will formulate it correctly?
>
> 4. We feel that we have free will, that we are free to choose whether to do something or not. But if everything is predetermined by scientific laws, then free will is an illusion. What is our responsibility for our actions based on? We do not punish criminals if they are insane, because we believe that this will not help.

In the aforementioned article, "The Clockwork Universe in Chaos," and in several of his other works, Richard Thompson suggests that God would be able exert control over the course of world events, without seemingly breaking known physical laws, by operating at the micro (quantum) level, which is below our threshold of perception. Blaise Pascal, 17th century "father of probability theory," is credited with the statement: "Chance is another name for God." Quantum mechanics establishes limits for microscales of space and time (Planck length, Planck time, etc.), as formulated in the famous Heisenberg uncertainty principle. It is at this level, elusive for us, that God is able to exercise control over the universe. As Albert Einstein said, "God does not play dice."

Another objection to the philosophy of determinism is the experience of free will by all living beings, as mentioned by Hawking in point 4 above. Attempts to explain the very phenomenon of consciousness, along with everything that happens to each individual, from the point of view of mechanistic principles certainly appears to deprive living beings of freedom to make meaningful choices. One of the more famous attempts to formulate a theory that allows for both free will and determinism is Hugh Everett's many-worlds interpretation of quantum mechanics. According to this theory, if a person is making a choice whether to go to the right or to the left, then two universes automatically arise, one in which he goes to the right, and the other in which he goes to the left. Thus, we get an infinite set of universes, or an infinite bifurcation of space-time. This theory, however, does not have a large number of adherents due to its artificial nature and the impossibility of its experimental verification.

The Vedic scriptures do not deny some amount of predestination, caused by the law of *karma*. Nevertheless, every living entity has what may be called a "corridor of *karma*," where the "width" depends on the degree of self-realization (level of consciousness) of the living entity and how he is being influenced the *guṇas* (modes of nature: goodness, passion, and ignorance). An illustrative analogy in this regard is given in the commentary to verse 11.3.6 of *Śrīmad-Bhāgavatam*:

> If I purchase a ticket for an airline flight, board the plane and commence the flight, once the plane has taken off my decision to board the plane forces me to continue flying until the plane lands. But although I am forced to accept the reaction of this decision, on board the plane I have many new decisions I can make. I may accept the food and drink from the stewardesses or reject it, I may read a magazine or newspaper, I may sleep, walk up and down the aisle, converse with other passengers, and so on. In other words, although the general context – flying to a particular city – is forcibly imposed upon me as a reaction to my previous decision to board the plane, even within that situation I am constantly making new decisions and creating

> new reactions. For example, if I cause a disturbance on the airplane I may be arrested when the plane lands. On the other hand, if I make friends with a businessman sitting next to me on the plane, such a contact may lead to a favorable business transaction in the future.
>
> Similarly, although the living entity is forced to accept a particular body by the laws of karma, within the human form of life there is always scope for free will and decision-making. Therefore the Supreme Personality of Godhead cannot be considered unjust for holding the living entity in human life responsible for his present activities despite the living entity's undergoing the reactions of his previous work.

Another illustration of the possibility of going beyond the bounds of predestination is the well-known story in connection with the Buddha. Once a famous astrologer from Kāśī was walking along the river bank and saw footprints in the sand. Looking at the footprints, the astrologer was perplexed, because, according to his books, these footprints must have belonged to the ruler of the universe. However, the footprints led the astrologer to a tree, under which a beggar in rags was sitting. It was Buddha. Approaching him, the astrologer was even more confused – he saw a beggar, but he understood that this man was essentially a king. He asked the Buddha to dispel the doubts.

Buddha opened His eyes and said,

> What the books have taught you will be true in the case of hundreds of thousands of other people. But now you have met the one to which your books are not applicable. By being mindful, I don't make the same mistake twice. Nobody can predict the next moment in my life.

As Jay Israel (Jayādvaita Swami) points out in his book, *Vanity Karma: Ecclesiastes, the Bhagavad-gītā, and the meaning of life*, almost everything in a person's life is predetermined, except for one thing – how they react to the words of sages who urge us to embark on the path of self-realization.

Conclusions

The Vedic scriptures interpret time as the energy that drives all the processes of creation, maintenance, and destruction of the material world. This approach does not contradict the ideas of modern science, but rather complements and harmonizes its ideas about the nature and measurement of time. What practical conclusions can we draw for ourselves? According to the principle of *yukta-vairāgya*, one should engage this energy, time, in service to the Source of all energy. In verse 2.3.17 of *Śrīmad-Bhāgavatam* it is said that with every sunrise and sunset, the sun shortens the life of everyone, except for those who dedicate their life to the service of the Supreme Personality of Godhead. As Cāṇakya Paṇḍita said in the *Nīti-śāstra* (civic laws): *āyuṣaḥ kṣaṇa eko 'pi na labhyaḥ svarna-koṭibhiḥ* (even one moment of life cannot be returned even for millions of gold coins. What could be a greater loss than wasted time?) This is confirmed in the *Śrīmad-Bhāgavatam* (verse 2.1.12):

kiṁ pramattasya bahubhiḥ
parokṣair hāyanair iha
varaṁ muhūrtaṁ viditaṁ
ghaṭate śreyase yataḥ

> "What is the value of a prolonged life which is wasted, inexperienced by years in this world? Better a moment of full consciousness, because that gives one a start in searching after his supreme interest."

As an example of the correct use of time in the *Śrīmad-Bhāgavatam*, the following verse (2.1.13) gives the example of Mahārāja Khaṭvāṅga:

khaṭvāṅgo nāma rājarṣir
jñātveyattām ihāyuṣaḥ
muhūrtāt sarvam utsṛjya
gatavān abhayaṁ harim

> "The saintly King Khaṭvāṅga, after being informed that the duration of his life would be only a moment more, at once freed himself from all material activities and took shelter of the supreme safety, the Personality of Godhead."

In a purport to this verse, Bhaktivedanta Swami Prabhupāda writes the following:

> Mahārāja Khaṭvāṅga was invited by the demigods in the higher planets to fight demons, and as a king he fought the battles to the full satisfaction of the demigods. The demigods, being fully satisfied with him, wanted to give him some benediction for material enjoyment, but Mahārāja Khaṭvāṅga, being very much alert to his prime duty, inquired from the demigods about his remaining duration of life. This means that he was not as anxious to accumulate some material benediction from the demigods as he was to prepare himself for the next life. He was informed by the demigods, however, that his life would last only a moment longer. The king at once left the heavenly kingdom, which is always full of material enjoyment of the highest standard, and coming down to this earth, took ultimate shelter of the all-safe Personality of Godhead. He was successful in his great attempt and achieved liberation. This attempt, even for a moment, by the saintly king, was successful because he was always alert to his prime duty...

PART THREE

Artificial Intelligence: Can a Computer Possess Consciousness?

Artificial Intelligence: Can a Computer Possess Consciousness?

"Simulating the activity of consciousness inside a computer is like modeling weather patterns: you can simulate what happens in the weather world, but you don't have clouds and winds inside the computer."
Professor Anil Seth

Throughout its history, mankind has created various mechanical devices to facilitate physical labor. Similarly, the degradation of human mental abilities during the Kali-yuga was one of the reasons for the technological progress towards the facilitation of mental activity. The abacus (accounts) used in various ancient civilizations were gradually replaced by arithmeters,[1] the first of which is believed to have been created by Leonardo da Vinci around 1500.

Figure 1: Ada Lovelace (1818–1852), English mathematician and writer, chiefly known for her work on Charles Babbage's proposed mechanical general-purpose computer, the Analytical Engine.

Interestingly, Ada Lovelace, daughter of the famous English poet Lord Byron, is considered to be the creator of the first public computer. Then, in 1843, Ada (whose name is used for a well-known programming language) noticed that the "Analytical Engine" invented by her had no pretensions to create anything and could do "only what a person orders it to do, and the task of a

1 A mechanical contrivance for multiplication and division, based on the logarithmic principle, a form of cylindrical slide rule.

computing machine is to help us make accessible what we are already familiar with."[2]

However, in his landmark 1950 article, "Computing and Intelligence," the famous English inventor Alan Turing concluded that general-purpose computers could be capable of learning and originality, thus formulating the key points of the concept now known as Artificial Intelligence (AI). Turing believed that a computer is capable of mimicking any human action and humans are ultimately machines. Gradually, in the second half of the 20th century, an opinion was formed that within a few centuries, or perhaps even decades, computers would be fully equal in capability to the human brain, and "personal" biorobots from the novels of Isaac Asimov, Douglas Adams, or George Lucas would become full-fledged members of human society.

Figure 2. Alan Turing (1912–54), English mathematician, computer scientist, logician, cryptanalyst, philosopher, and theoretical biologist along with his 1936 "Turing Machine."

The concept of "Artificial Intelligence" was originally introduced by the famous scientist from Stanford University, John McCarthy, in 1956 at a conference at Dartmouth University.[3] It should be noted that McCarthy's conception made no pretense of reproducing human

2 See: *Scientific Memoirs*, edited by Richard Taylor (1781–1858), Volume 3, "Sketch of the Analytical Engine invented by Charles Babbage, Esq," Notes by the Translator, by Augusta Ada Lovelace (1843). https://en.wikipedia.org/wiki/Computing_Machinery_and_Intelligence.

3 The 1956 Dartmouth Summer Research Project on Artificial Intelligence was widely considered to be the founding event of artificial intelligence as a field.

intellectual capabilities. For him, the abilities of computer programs correspond to the intellectual mechanisms that program designers understand well enough to put in programs.

No one is surprised by the ability of physical mechanisms to perform physical work that is beyond human capacity. So why should we be surprised by the ability of computer programs to lighten the intellectual load of their human creators? According to McCarthy, "The problem is that we cannot yet generally define what computational procedures we want to call 'intelligent.' We understand some mechanisms of intelligence and do not understand the rest. Therefore, intelligence within this science refers only to the computational component of the ability to achieve goals in the world."[4]

The "cooling" of the ardor of those wishing to build artificial intelligence systems that fully replicate human thinking in the following decades led to the emergence of two narrower concepts: "strong artificial intelligence" and "weak artificial intelligence."

The theory of strong artificial intelligence suggests that computers may acquire the ability to think and realize themselves as individuals (in particular, to understand their own thoughts), although it is not necessarily the case that their thought process will be similar to that of humans.

On the contrary, the theory of weak artificial intelligence rejects this possibility and implies the creation of multifunctional systems that replicate (or surpass) human intelligence for a specific purpose. A typical popular example of a weak artificial intelligence system is software for speech synthesis or recognition, unmanned driving, voice-activated web search, or playing chess.

For instance, the world community was shocked by the 1997 loss of World Chess Champion Garry Kasparov to IBM's Deep Blue computer system. Having originally accused IBM of fraud, Kasparov later revised his attitude and began to study in detail the scientific and philosophical problems of artificial intelligence systems. His shifting views on this issue are reflected in the book, *Man and Computer: A Look into the Future*.

According to Kasparov, the understanding of artificial intelligence,

4 J. McCarthy, "What is Artificial Intelligence?" (1997). Available electronically at http://www-formal.stanford.edu/jmc/whatisai/whatisai.html.

Figure 3. Garry Kasparov (born 1963), Russian chess grandmaster, former World Chess Champion (1985–2000), political activist, and writer.

which is fundamental to computer science, is problematic. The basic assumption behind Alan Turing's vision for artificial intelligence is that the human brain is so similar to a computer that we should be able to create a machine that successfully imitates human behavior. But despite the elegance of the classic analogy – neurons as transistors, cortex as a memory bank, etc. – we find scant biological evidence supporting this claim, and it tends to obscure fundamental dissimilarities between human and machine "thinking." The infamous quote attributed to Pablo Picasso, "Computers are useless because they can only provide answers,"[5] highlights a crucial difference between mechanistic and nonmechanistic "intelligence": An answer implies an end, a full stop, but for Picasso, and indeed for many first-rate thinkers, there is no end to inquiry – only more and more questions.

Traditional Western science generally considers the brain to be the source of both consciousness and intelligence, and indeed, often conflates the two. On the contrary, according to the *Bhagavad-gītā,* intelligence is a subtle *material* element that interacts with the brain. Above the gross senses, the brain, and the subtle intelligence is the soul, the source of consciousness, which belongs to the spiritual energy: "the spirit soul is not the body."[6] This sacred text explains that the interaction of the human intellect with the brain, as well as with the soul, is under the control of Paramātmā, the Supersoul, or the all-pervading universal consciousness. This offers an alternate understanding to the common scientific position of intellect as a product of brain neuron activity.

5 Fred R. Shapiro, *The Yale Book of Quotations,* Section Pablo Picasso (New Haven, CT: Yale University Press, 2006), 591.

6 A. C. Bhaktivedanta Swami Prabhupāda, *Bhagavad-gītā As It Is,* 3.41, purport (Los Angeles, CA: The Bhaktivedanta Book Trust, 1994), 209.

Moreover, a well-known expert in the field of artificial intelligence, Professor Andrew Yan-Tak Ng of Stanford University, believes that artificial neural networks built by deep learning engineers are relatively simple and cannot claim to reproduce the activities of biological neurons in the brain. According to this authoritative expert, we have almost no idea how the brain works, and the answer to the fundamental question of how exactly a neuron transforms information is unlikely to be answered in the near future.

Figure 4. **Andrew Yan-Tak Ng (born 1976), British-American computer scientist and technology entrepreneur focusing on machine learning and artificial intelligence.**

Formally, the term "neural network" refers to neurobiological processes. But although some of the central concepts of deep learning have been developed, in part, based on our understanding of the brain, deep learning models are not necessarily related to brain activity (even assuming that the brain is the source of consciousness and mental activity). There is no evidence that the brain implements anything like the learning mechanisms used in current neural network models of deep learning. As an authority in this field, Andrew Ng says, "You may come across popular science articles claiming that deep learning works like the brain or has been modeled after the brain, but this is not true. It will be confusing and counterproductive for beginners."[7]

Another world-famous personality, Nobel laureate in physics Sir Roger Penrose, also takes a tough stance concerning the impossibility of explaining consciousness using computer models. "Computers can do amazing things – play chess, etc. But the ability to *understand* comes from the individual who programs the computer. The computer machine itself does not possess the quality of understanding."[8] To justify his point of view, Penrose refers to Gödel's famous incompleteness theorem, which says that in any formal theory there are unprovable statements. The point of Penrose's reasoning is that when faced with a

7 Deeplearning.ai "Advanced Learning Algorithms" course, week 1.

8 Sir Roger Penrose, *Shadows of the Mind: A Search for the Missing Science of Consciousness* (New York: Oxford University Press; First edition, 1994).

logically unsolvable problem, the human mind easily adds to the system of axioms to solve it, i.e., goes beyond the axioms while remaining fixed on the truth. However, the most advanced computer is not capable of this, unless the very mechanism of forming new hypotheses is described for it by a human.

Figure 5. Sir Roger Penrose, (born 1931), British mathematical physicist, philosopher of science, and Nobel Laureate in Physics. He is Emeritus Rouse Ball Professor of Mathematics in the University of Oxford, an emeritus fellow of Wadham College, Oxford, and an honorary fellow of St John's College, Cambridge, and University College London.

To assess the ability of computers to reproduce the properties of human intelligence, an evaluation procedure called the "Turing test" was introduced in 1950 by Alan Turing. Turing's goal was to determine whether a machine could think. The essence of the test is as follows: a person interacts with unknown entities, and based on the answers to his questions, he must determine whether he is talking to a human or a computer program. If our agent cannot say for sure which of the interlocutors is a human, the machine is considered to have passed the test.

In 2022, Google's LaMDA and ChatGPT AI systems became the first to pass the Turing test. Nevertheless, many of today's leading scientists are skeptical that the Turing test is adequate for evaluating the thinking abilities of computing systems.

The scientific community is familiar with John Searle's thought experiment known as the "Chinese Room." The purpose of the experiment is to refute the claim that a digital machine endowed with "artificial intelligence" is capable of possessing consciousness in the same sense in which humans possess it. In other words, the goal is to refute the hypothesis of so-called "strong" artificial intelligence and thus discredit the Turing test. The essence of the experiment is as follows. Let us imagine an isolated room containing a person who does not know a single Chinese character. However, he has precise instructions written down in a book on how to manipulate the characters, such as "Take this character from basket number one and place it next to this character from basket number two." But these instructions lack information about the meaning of the characters, so he simply follows these instructions like a computer. An observer who knows Chinese characters passes a question written in Chinese into the room and expects to get a conscious answer at the end.

Figure 6. John Searle's "Chinese Room" thought experiment, illustrating that a computer cannot be "conscious," regardless of how human-like it may seem.

The instruction is designed in such a way that after applying all the steps to the characters of the question, they are transformed into the characters of the answer. In fact, these instructions are similar to a computer algorithm, and a person executes the algorithm in the same way as a computer would execute it. At the same time, the person himself does not have any knowledge of hieroglyphs and does not understand either the original question or the answer, which he himself compiled. An observer, however, may conclude that there is a person in the room who knows and understands the hieroglyphs. Thus Searle concludes that although such a system can pass the Turing test, no understanding of language occurs within the system, and thus the Turing test is not an adequate judge of thinking ability. Searle's arguments are aimed at criticizing the position of so-called "strong" artificial intelligence, according to which computers with the appropriate program can actually understand natural language, as well as possess other mental abilities peculiar to humans.

Discussing Searle's experiment, Penrose concludes that the claims not only of "strong" but also of "weak" artificial intelligence are overstated. And he further suggests that, in general, no computer program is capable of simulating manifestations of consciousness. To explain consciousness requires an appeal to something "genuinely incalculable,"[9] which means, Penrose concludes, that we must look for an appropriate place where a major gap exists in the scientific picture.

To summarize, we note that significant successes in the development of weak AI systems do not indicate any progress in the field of building strong AI systems. Moreover, the very prospect of building strong AI systems is seen by many authorities in this field as epistemologically impossible. Human and even animal behavior cannot be algorithmized, as it depends on the ability to directly experience or intuit the truth. In Vedic philosophy, this is called *aparokṣa* – direct acquisition of knowledge, bypassing the level of feelings and mind. Emotional and, moreover, spiritual experience by its nature is not subject to algorithmization, although *some* external aspects of the manifestation of consciousness can be described and reproduced as a result of modeling.

So even though human thinking has a formal-logical component, it is fundamentally not subject to algorithmization (what to speak of the possibility of computers acquiring transcendental knowledge that is outside the sphere of action of the modes of material nature). In one sense, the brain may be compared to a computer equipped with software of subtle mind and intelligence, but its user is the soul. The brain, like a computer, acts on the basis of certain logical rules, but these rules are established by the Supersoul, who remains outside the system.

Pleasure and pain, perception of beauty and sense of humor, consciousness and free will – will any of these abilities arise in electronic robots, even if the algorithms that control them acquire a "sufficient

---

9 Penrose, *Shadows of the Mind.*

degree of sophistication"? As Penrose writes in *Shadows of the Mind*, "No physical, biological, or mathematical theory has yet come close to explaining consciousness and its logical corollary, intelligence."[10]

As mentioned above, no one is surprised by the ability of physical mechanisms to out-perform humans on physical tasks. So why should we be surprised by the ability of computer programs to lighten the intellectual load of their creator? The agent in either case is a living being, the soul, who is transcendental to both the gross (physical) and the subtle (mental, intellectual) body.

Figure 7. ***Bhagavad-gītā As It Is*****, a translation and commentary of the *****Bhagavad-gītā*** **by A. C. Bhaktivedanta Swami Prabhupadā, founder of the International Society for Krishna Consciousness.**

Vedic scriptures often use the term *māyā*, or illusion, in reference to our futile attempts to understand reality by measuring all that exists: *nāhaṁ prakāśaḥ sarvasya, yoga-māyā-samāvṛtaḥ* ("I am never manifest to the foolish and unintelligent. For them I am covered by My eternal creative potency").[11] Undoubtedly, modern computer technology cleverly harnesses electrical energy to power devices that are able to manipulate symbols at tremendous speed, allowing them to perform extremely cumbersome and complex tasks. But can we expect a computer to be actually "conscious" of its actions, any more than we expect a calculator or an ordinary wooden hammer to be? We will leave it to the reader to answer this question.

10 Penrose, *Shadows of the Mind*.

11 Prabhupāda, *Bhagavad-gītā* 7.25, 399–400.

www.ingramcontent.com/pod-product-compliance
Lightning Source LLC
Chambersburg PA
CBHW070449050426
42451CB00015B/3403
9781959829058